SAHARAN WINDS

ENERGY AND SOCIETY
Brian Black, Series Editor

American Energy Cinema
Edited by Robert Lifset, Raechel Lutz, and Sarah Stanford-McIntyre

Engaging the Atom: The History of Nuclear Energy and Society in Europe from the 1950s to the Present
Edited by Arne Kaijser, Markku Lehtonen, Jan-Henrik Meyer, and Mar Rubio-Varas

Transportation and the Culture of Climate Change: Accelerating Ride to Global Crisis
Edited by Tatiana Prorokova-Konrad

Energy Culture: Art and Theory on Oil and Beyond
Edited by Imre Szeman and Jeff Diamanti

On Petrocultures: Globalization, Culture, and Energy
Imre Szeman

Oil and Urbanization on the Pacific Coast: Ralph Bramel Lloyd and the Shaping of the Urban West
Michael R. Adamson

Oil and Nation: A History of Bolivia's Petroleum Sector
Stephen C. Cote

Saharan Winds

Energy Systems and Aeolian Imaginaries in Western Sahara

Joanna Allan

WEST VIRGINIA UNIVERSITY PRESS / MORGANTOWN

First edition published 2024 by West Virginia University Press
Printed in the United States of America

ISBN 978-1-959000-23-5 (paperback) / 978-1-959000-24-2 (ebook)

Library of Congress Cataloging-in-Publication Data
Names: Allan, Joanna, author.
Title: Saharan winds : energy systems and aeolian imaginaries in
Western Sahara / Joanna Allan.
Other titles: Energy and society.
Description: Morgantown : West Virginia University Press, 2024. |
Series: Energy and society | Includes bibliographical references and index. |
Identifiers: LCCN 2024009344 | ISBN 9781959000235 (paperback) |
ISBN 9781959000242 (ebook)
Subjects: LCSH: Energy development—Political aspects—Western Sahara. |
Winds—Western Sahara. | Wind power—Social aspects—Western Sahara. |
Winds in literature. | Western Sahara—Colonization. | Western Sahara—History.
Classification: LCC HD9502.A35542 A553 2024 |
DDC 333.79150948—dc23/eng/20240228
LC record available at https://lccn.loc.gov/2024009344

Book and cover design by Than Saffel / WVU Press
Cover image: Dusty. Western Sahara/Flickr, CC BY-SA

To Sandro, Logan, and Helios

Contents

Acknowledgments

This book would not have been possible without the support of numerous people and institutions. First, I profoundly thank the Leverhulme Trust for the Early Career Fellowship that funded the research presented in this book. Furthermore, the trust's flexibility and the helpfulness of its grant officers made it easier to have two children while carrying out the research.

I am in debt to all at West Virginia University Press for making this book possible. A special thanks to Derek Krissoff for his belief in the book; to Michelle Scott and the team at Westchester Publishing Services for the wonderful copyediting; and to the Energy and Society series editor, professor Brian Black, for his support and advice on the book and beyond. Thank you to Jo Bettles for her wonderful work on the index. A huge thanks to the peer reviewers, one anonymous and the other Alexander Dunlap, who generously dedicated much time and energy to helping me improve the script.

I have been lucky to benefit from wonderful colleagues at all stages of my career. I thank my former colleagues in the Hispanic studies department and wider Modern Languages School at Durham University who supported me in applying for the fellowship that made this book possible, and later in reading drafts of a chapter of this book. In particular, I am grateful to Luke Sunderland for his support with my fellowship application, to Kerstin Oloff and Rosi Song for their mentorship during my time at Durham and since, and to Nick Roberts not only for reading drafts of my work but also for always sticking up for junior colleagues. I am also indebted to professor Jon Gluyas, director of Durham Energy Institute, for helping me to understand more about the LPG industry.

I have benefited from useful feedback to talks I have given while working on the book. Thank you to Awa Sarr, Kerstin Oloff, Stephanie Benazquen-Gautier, Judit Tavakoli, Caitríona Ní Dhúill, and Mattin Biglari for the invitations, as well as audiences at conferences hosted by the Association of Hispanists of Great Britain and Ireland, the Petrocultures research network, the Royal

Geographical Society, the African Studies in Europe research network, and the Global Extractivisms and Alternatives Initiative. Thanks also to staff at the Spanish General Administration Archives for all their help locating useful documentation.

I am grateful to my colleagues (past and present) and the supportive, nurturing environment at the Northumbria Centre for Global Development and the wider Geography and environmental sciences department. In particular, I thank Katy Jenkins and Matt Baillie-Smith for having enough faith in me to give me my first permanent academic post, for their ongoing support, mentorship, and friendship, and for making the Centre for Global Development such a lovely place to work. Thank you also to Northumbria University for the Vice Chancellor's Research Fellowship, which afforded me time to develop this book. Also at Northumbria, I thank Tanya Wyatt and Kate Maclean for their mentorship and Ariane Bogain for the support she has offered to me and to so many others. A special thanks to Katherine Baxter for her mentorship, generosity, and support, which has been enormously valuable to me, as well as for encouraging me to think in new ways.

There are several fellow researchers working currently, or in the past, in Western Sahara to whom I am grateful for various reasons, especially Raquel Ojeda, Nick Brooks, José Antonio Rodríguez Esteban, Mahmoud Lemaadel, Pablo San Martín, Bartek Sabela, Blanca Camps-Febrer, Benita Sampedro Vizcaya, Alice Wilson, Fico Guzman, Mark Drury, Sebastien Boulay, and Hamza Hamouchene. Teachers, mentors, and supervisors from my student days continue to be a great source of support to me, especially Manuel Barcia Paz, Thea Pitman, and Claire Taylor. Thanks also to Ray Bush for his support and to Hannah Kate Boast for her constructive readings of my work and her advice. Students at Durham and Northumbria have been a source of inspiration to me.

Thank you to the poets and translators who have allowed me to reproduce their beautiful work in this book.

I am grateful to fellow volunteers at Western Sahara Campaign UK, Western Sahara Resource Watch, and other solidarity campaigns—especially John Gurr, Beccy Allen, Sara Eyckmans, Eric Hagen, and Asria Mohammed Taleb—who still inspire me, energize me, and teach me even though I have not seen some of them as much as I did pre-COVID and pre-children. Thank you to the Polisario Front, especially its permanent representative to the UK Sidi Brieka, for always facilitating my ability to carry out research in the Saharawi state-in-exile and allowing me complete freedom in my research activities there. Thanks also to the Polisario Front for granting me permission to carry out research in occupied Western Sahara and to the Algerian embassy in London for permitting me to carry out research in Algeria.

This book would have been impossible without the assistance—practical and intellectual—of Nguia Mohammed, Fadili Mohammed, Khaira Aayache, and my dear friend and research collaborator Moiti Mohammed. I am indebted to many other Saharawis for their kindness, assistance, and teachings, but it is prudent that they remain anonymous here. Mahmoud Lemaadel and Hamza Lakhal have kindly permitted me to edit, update, and include here, as a chapter, an article that we researched together. Thanks too to Hamza for his help with translations and for checking my transliterations.

I have known members of the Lakhal family—cut in two by the Wall of Shame—since 2006, when I first visited the Saharawi camps. None of my work in Western Sahara would be possible without them, and our friendships have shaped my life. A special thanks to Ambassador Malainin Lakhal and his family in the camps for support and encouragement over the years. And, on the other side of the wall, to my long-term research collaborator and dear friend Hamza Lakhal and all his family there, thanks are not enough.

I would be unable to carry out research without my family support network. Finally then, my utmost gratitude goes to my sister, Frances; my parents, Judith and David; my children; my uncle, Ian Williamson; my in-laws; and, above all, my partner, Helios Molinero Rodriguez.

Any errors in the book are mine alone.

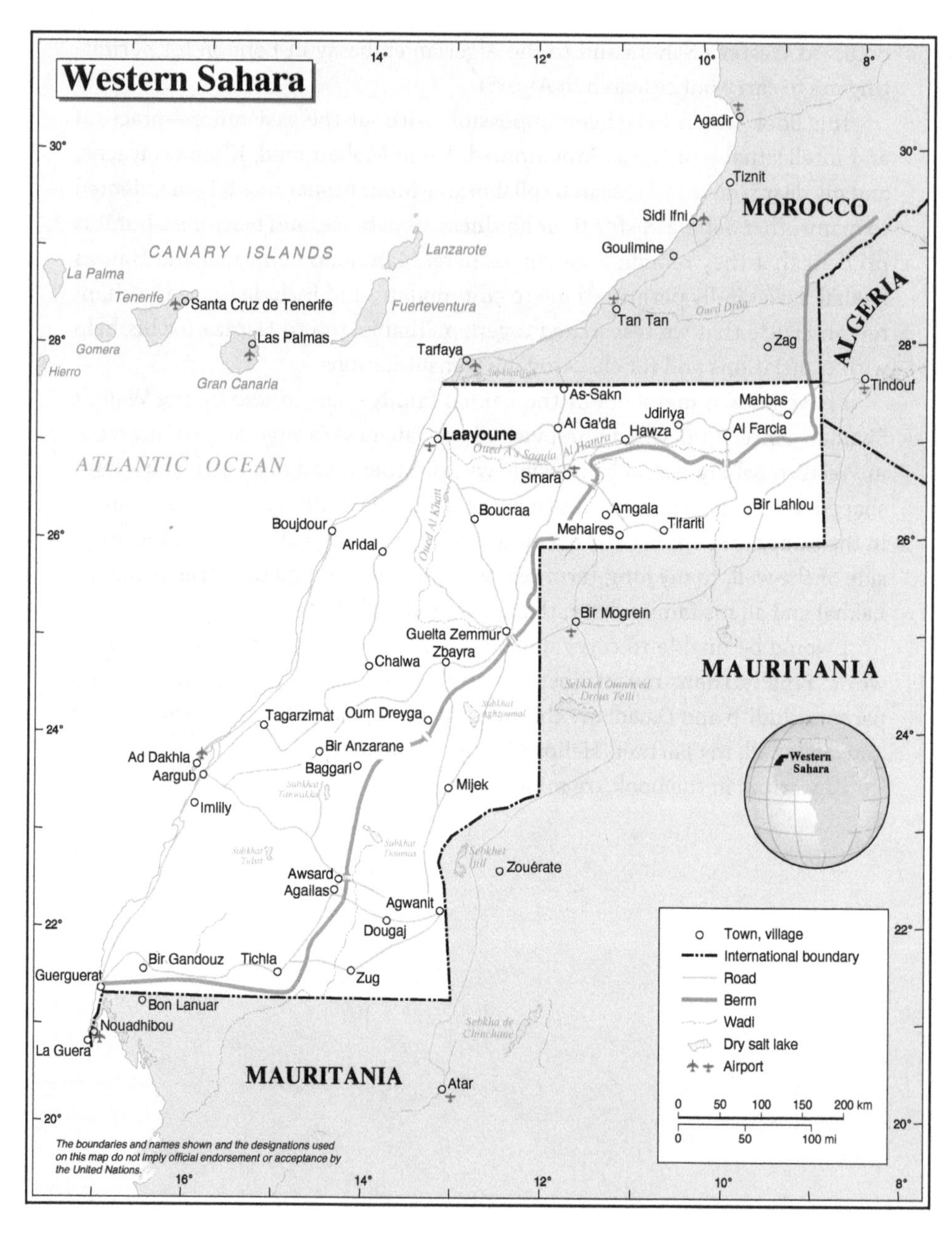

Western Sahara, map no. 3175, rev. October 4, 2012 (United Nations)

Introduction

The Helm

At Cross Fell, the highest summit in the Pennine hills, blows the Helm, a wind so strong that sheep have been known to dance in the air in its presence.[1] The Helm is one of the few named moving airs in England. Its naming is testament to the strength of its identity, or at least to the identity that my local community has given it over the centuries. As it descends the hills, the wind grows and unfurls clouds. These rest, stationary yet whirling, above Cross Fell, a gray-white helmet for the land and rivers below.

The Tees rises below the Helm. From a lichen-covered stone that marks its source, the river Tees smashes eastward through rocks and dales, broadens at Middleton, and then flows past Middlesbrough, a town nicknamed *Infant Hercules* for its role at the heart of the industrial revolution. The river oozes through mudflats and then eventually emerges at Teesport, land of oil, gas, petrochemical industries, toxic dumping, and ecological disaster. From here, the dying, polluted estuary carries international vessels, with multicolor shipping containers playing Tetris on their backs, out to sea, toward the Redcar offshore windfarm and beyond.

I have taken my boys ship spotting at Teesside port. A *freeport*. Free of the need to pay taxes on goods exported abroad. The Tees and its heavy industry, like all North East England, are historically tied up with industrial revolution, maritime empire, and an asymmetric contribution to the climate crisis. As my youngest boy points and oohs at ships from his pushchair and the older one observes them from his fluorescent orange toy binoculars, I am looking through the drizzle for a particular boat. A colleague has advised me that a vessel probably bound for the particular *beyond* that interests me is due to arrive at the port, a vessel and *beyond* that reveals Teesport's allegedly historic role in colonialism to be current.

Some days later, that boat arrives at El Aaiún port, in Moroccan-occupied Western Sahara, carrying liquified butane gas (LPG) sourced from the North Sea. One cannot be sure exactly where the LPG is destined for from here, but experience tells us probably the Moroccan army, the vast rows of refrigerators in the fisheries processing factories, the occupation's critical infrastructure, and its military wall.[2] In 2020, the UK became occupied Western Sahara's main exporter of LPG.[3] The exporting port for most of these shipments was my local Teesside. Few talk about Western Sahara in North East England, but marine paths join the dots between our peoples.

South from El Aaiún port, sleepless Abdelhay Layachi[4] pulls a blanket over his shoulders. An icy draft has come off the sea and sneaks into the tent. He is camping at the Guerguerat roadblock once again. Layachi has a long CV of participation in protest camps, but something in the air tonight has made him feel different, expectant, and uneasy. Guerguerat is Western Sahara's last stopping place on the modern trans-Saharan route between northern Morocco and the Senegal River.[5] This place has several nicknames. To Saharawis, it is *the plunder corridor*, the place through which their resources are stolen and transported to buyers elsewhere. More widely, as Mark Drury notes, this place is nicknamed *Kandahar* after a city in Afghanistan that came to symbolize the ungovernable, or "no man's land."[6] Guerguerat sits where three borders meet. At the roadside, it is still a while until dawn, but Layachi senses lights. There is clattering, raised voices, and the hum of vehicles. He gets up and takes his phone. Before he is out of the tent, there is an agonizing creak, like the stress of matter moving over matter, or a ship leaning forward to sink. Something has breached the wall.

Guerguerat village is in the Moroccan-occupied part of Western Sahara. However, the roadblocks that Layachi and his friends have mounted on several occasions since 2016 have been down the road in Polisario-controlled Western Sahara. The Polisario Front is the nationalist liberation movement of the Saharawi people. The Polisario controls a twenty-five-kilometer-long strip of land between the Moroccan-occupied zone and the Mauritanian border (as well as around a third of Western Sahara to the east), which the UN designated as a demilitarized buffer zone as part of a 1991 cease-fire agreement between Polisario and Morocco.

On November 13, 2020, Layachi became a witness to the end of that twenty-nine-year cease-fire. "It was about 6:00 in the morning," he tells me on a WhatsApp voice note.[7] "The Moroccan military had made a breach in the military wall. I saw at least two vehicles come through this breach, driving towards us, to attack us. Then they opened fire." Polisario responded to the Moroccan military by firing back, Layachi says, and then evacuated him and his colleagues and drove them to the Saharawi refugee camps near Tindouf, western Algeria. Polisario's interpretation of events chimes with that of Layachi. As Sidi Breika, the Polisario representative to the UK, explained to me, "Polisario Front decided to withdraw from the cease-fire agreement as a response to the military aggression of Moroccan army against Saharawi civilians who were legally and peacefully protesting against the Guerguerat illegal breach and the continuous plundering of their natural resources."[8]

On hearing of the outbreak of war, I worry about Saharawi friends. I call them or message them if they do not pick up. One friend, Hamza, after many months of absence, is suddenly online again. He is rapidly writing and publishing poetry in response to the developments. I help him translate the poems he is posting to Facebook. He hopes they will reach a wider audience if they exist in English as well as Arabic.

Outcry!

An outcry. A jagged outcry,
From the cities of revolutions,
From my homeland,
From the desert's anger,
From the rhythmic pulse of my wound,
Which beats through a rush of songs.

An outcry,
Asking the wind about weeping.
Women wailing
In the midst of a silence
That almost fills this world with noise!

An outcry
Asking humanity
About a prisoner still dragged by history
Deafeningly
Into the depths of eternity![9]

صرْخَةٌ

صرْخَةٌ مِنْ مَدَائِنِ الثوَرَاتِ
مِنْ بِلَادِي..
مِنْ غَضْبَةِ البِيدِ،

مِنْ إيقَاعِ جُرْحِي فِي زَحْمَةِ الأغْنِيَاتِ
تَسْأَلُ الرِّيحَ عَنْ نُوَاحِ عَبَارَى
وَسْطَ صَمْتٍ
يَكَادُ يملأ هَذِي الأرْضَ ضَجًّا!
تُسَائِلُ البَشَرِيَّهْ
عَنْ سَجِينٍ مَا زَالَ يَسْحَبُهُ التَّارِيخُ غَصْبًا
فِي غَيْهَبِ الأبَدِيَّهْ

I read the poem online a fortnight after war breaks out. Hamza first published it with the dedication "This poem is an outcry for our heroes 'The Gdeim Izik Political Prisoners.'" These prisoners are a group of men serving long sentences for their alleged role in organizing the largest protests ever seen in Western Sahara. Their mothers and sisters—"Women wailing"?—are among the loudest advocates for their release. Do I understand Hamza's meaning fully? To help me translate, I text Hamza to ask what he wishes to convey with the poem. His reply reads, "So this outcry asks the wind about these weeping women, who have been asking for help for many years. And the other meaning too, is that the wind bears sound and takes it everywhere. So, it's like asking the wind 'why in all these years has nothing changed? Did you not send our message? Or did the others not want to hear?'"[10] Hamza makes the wind a messenger but also an agent with whom humans can communicate political agendas. Hamza also points to the properties of wind and its role in transmitting sound. Wind helps to express his reaction to this new war, which was triggered by resource plunder.

The links between corporate complicity in conflicts, neocolonialism, and human rights are a long-term interest of mine. Before embarking on an academic career, I worked at nongovernmental organizations (NGOs) that campaigned against such corporate activities. In my free time, I volunteer with two NGOs that focus specifically on corporations, as well as states and groups of states such as the EU, implicated in the plunder of Western Sahara. As the attention of these NGOs turned a few years ago to wind energy infrastructural developments in occupied Western Sahara, my *hobby*, if that is the right word, began to converge with my academic training in cultural studies. I wondered how the energy corporations that we lobbied against imagined the Saharan lands, waters, winds, and sun that they wished to exploit. How did their imagined resources differ from how Saharawis imagined or understood the same nonhuman elements of the desert? For Hamza, a young man of the urban era but familiar with the traditional Saharawi poetry of nomads, Saharan winds serve as metaphors. For his late father, who was a nomad, the winds had practical uses and meanings as well as poetic ones. Nomads interviewed for this project summed up the role of the winds in their lives variously as "a gift from God," "the engine of Saharawi human beings' relationship with the rest of nature," and "how we predict our futures and plan our days."[11] How do multinational energy corporations imagine the winds they (in their words) *harness* in occupied Western Sahara? To what extent do their imaginations affect how energy systems are planned, built, and managed? And do the corporations draw on older European colonial readings of the wind?

My project is shaped by such questions. Sun heats, earth moves, and wind is born. Winds are ubiquitous. But how we—as individuals, communities, or cultural groups—interpret winds varies greatly. Our interpretations of wind, or *wind imaginaries*, have political consequences and, more specifically, political consequences for our energy systems, renewable or otherwise. Western Sahara is the best case through which to elaborate this argument because, one of the windiest places in Africa, it is today a wind energy powerhouse enmeshed in an ongoing anti-colonial war. This conflict has its roots in the age of Europe's colonization of Africa; thus, my study of Saharan wind imaginaries, and their political effects, spans over two centuries.

Western Sahara is Africa's last colony, or, to use UN jargon, the continent's only remaining *non-self-governing territory* on the UN's list for decolonization. Spain acquired Western Sahara (as well as Equatorial Guinea), which became known to Europeans as *Spanish Sahara*, at the infamous 1884 Berlin Conference. At this meeting, which saw the colonial partition of Africa, the European powers used a ruler and pencil to divide the continent among themselves. The Spanish sought to profit from their African colonies' resources. In Spanish Sahara, they built a mine to extract phosphates and developed infrastructure to service a potential oil industry. In the late 1960s, by which time most African countries had achieved independence, Spain was under pressure from the UN and the Saharawi nationalist movement to decolonize. It never did so. Rather, it forged a secret agreement, illegal under international law, with two of Western Sahara's adjacent neighbors: Morocco and Mauritania. The terms of this agreement gave the two North African territories a portion of Western Sahara, while Spain would retain the right to fish in the latter's coastal waters and receive a 33 percent share of annual profits from the phosphates mine. Spain withdrew most of its citizens by October 1975, making way for the Mauritanian and Moroccan armies to invade. The International Court of Justice issued an advisory opinion in the same month, which concluded that there was not "any tie of territorial sovereignty between the territory of Western Sahara and the Kingdom of Morocco or the Mauritanian entity" and therefore supported "the principle of self-determination through the free and genuine expression of the will of the peoples of the Territory."[12]

Dozens of thousands of Indigenous Saharawi civilians fled eastward. The Moroccan air force bombed the fleeing refugees with napalm on four known occasions.[13] Others joined the Polisario Front, a pro-independence movement that had begun an armed struggle against Spanish colonialism in 1973. Algeria offered Saharawis asylum and space to build camps near Tindouf, a military outpost in its western corner. In February 1976, the Polisario declared these camps the Saharawi Arab Democratic Republic (SADR)-in-exile. Some 180,000 Saharawis live there today. A small population of Saharawis also remains in occupied Western Sahara, where they are vastly outnumbered by Moroccan settlers and subject to systematic human rights abuses. Around a third of Western Sahara, the *liberated territory* as Saharawis call it, is ruled by Polisario. The occupied region is separated from the liberated region by the longest active military wall—the wall Layachi saw breached in November 2020—in the world, built by Morocco in the 1980s.

In 1991, the UN brokered a cease-fire between Morocco and Polisario. The ceasefire agreement promised a referendum on self-determination, which included the option of independence, for the Indigenous Saharawis. However, Morocco, despite originally agreeing to the referendum, repeatedly blocked it.[14] The Saharawis waited for the international community to implement a diplomatic solution. And waited. And waited. In November 2020, they stopped waiting.

Morocco took advantage of the decades-long stalemate by, in violation of the Geneva Accords, moving a settler population into Western Sahara. It has also cashed in on its occupation by selling off Western Sahara's vast natural resources. Said resources take us back to Guerguerat village, the November 2020 end to a three-decade cease-fire, and the protests of Abdelhay Layachi and his friends. The roadblocks, according to Layachi, had one aim: "To close down the illegal breach at Guerguerat, [the] gate through which Morocco passes our plundered natural resources to Mauritania and other countries."[15] The new 2020 war is entangled with natural resource exploitation. Wind energy exploitation deserves special consideration when considering the role of resources in the Western Sahara conflict. This is because energy infrastructure makes Morocco dependent on Western Sahara for a portion of its electricity and, via pylons, cables, and so on, connects the two countries in a literal and material way.

With specific reference to the Western Saharan case, Mark Drury reminds us that so-called resources do not exist prior to the historical, sociological, and political processes that produce them as such.[16] Phosphates, oil, and indeed wind are imagined as, or understood to be, resources, only by specific communities in specific historical and political moments. Different cultures have not always understood the wind to be (only) a natural resource for producing energy. How cultures imagine, perceive, and understand the wind has political roots and political consequences.

My book offers a cultural history of the Western Saharan winds and how they have shaped energy politics. Such a history could be infinite. The winds that blow through Western Sahara and beyond have influenced cultural production across the globe for centuries. Take the Sirocco, a warm northeasterly wind that blows dust from the Sahara across the Mediterranean and, according to Europeans, agitates peoples' moods. It lends its name to a 1951 Humphrey Bogart movie, a 1987 album by the late flamenco guitarist Paco de Lucía, a bestselling 2016 cookbook of "fabulous flavours from the East" by Sabrina Ghayour, and a Volkswagen car.[17] No doubt countless other works are inspired by romantic dreams of the Sahara's airs. I therefore limit whose perceptions, and which political effects, I focus on. The borders of my exploration are guided by the battle that provoked my interest in Saharan winds in the first place: the (linked) anti-colonial, environmental, and climate justice struggles surrounding energy developments—renewable and otherwise—in occupied Western Sahara. I therefore focus on the wind imaginaries of the Saharawis, the Spanish colonizers, and colonizers from elsewhere in Europe and North America who washed up in the Sahara during the eighteenth- and nineteenth-century Age of Sail. In all these cases, and given my aforementioned interest in environmental, anti-colonial, climate, and energy struggles, I am interested in how wind imaginaries influenced energy developments in the territory that is today called *Western Sahara*. This is, then, a study of understandings and perceptions, but they are understandings and perceptions of tangible winds and windblown places that affected those places in specific ways.

How can humanities scholars and social scientists research and think through the wind? Weather shapes human experience. It is a cliché of British culture that small talk focuses on the weather. Yet, for a phenomenon that impacts our daily lives so greatly, there is scarce sociological, humanities, or political scholarship about the weather. Anthropologist Tim Ingold suggests this absence is down to Western ontology that understands life as existing on the outside of a solid sphere, on which the earth is the ground of our habitation, and which does not therefore allow for the phenomenon of the sky. As well as this, Ingold argues that the Western tendency to dichotomize phenomena into the categories of *material* or *immaterial* leaves us with little framework for understanding the weather. If the earth and sky are entirely separate, and we live on top of the earth and under the sky, then is weather part of the earth or sky? To overcome this conundrum, Ingold argues that earth and sky should be understood as "linked within one indivisible field," a "weather-world."[18] The world that we inhabit is an intermediate zone between the solid rock at the core of the earth and the sky above. In this world, the weather and land intermingle, and we live in this place of the crossover. For Ingold, this crossover, this weather-world, is a "medium" in which we live. We mingle with it and breathe it in and out.

If Western cultures have thought little about the weather, wind is especially forgotten. Rune Flikke suggests this is because Western ontology prioritizes visibility over other senses.[19] The favoring of the visible is linked to the history of colonialism. Says Leigh Eric Schmidt, the identification of the visual as supremely enlightened, Western, and modern has been sustained by "the othering of the auditory as 'primitive' or even 'African.'"[20] We can see the wind's physical effect on things and beings, but we cannot see the wind itself. For Flikke, wind's invisibility explains the West's tendency to ignore it. If we cannot see the wind, how do we know it is there, describe it, or read it? We might occasionally smell or even taste what is carried by wind, but we do not encounter the wind itself through these senses. We think we hear wind rustling, whistling, humming, and whooshing. But do we hear the wind itself, or rather the friction it produces?

The sense through which we most directly encounter the wind is touch. We feel wind's temperature, and we feel wind's pressure. In some places, we might feel what wind carries: dust in our eyes or wetness on our skin. Depending on what a strong wind hurls at us, we might also feel pain or loss of balance. Making wind the analytical space in which to think about the colonization of an African territory therefore forces us to consider the touchscape and the soundscape, as well as the landscape. As Nadera Shalhoub-Kevorkian says of the Israeli occupation of Palestine, an occupation is never just an occupation of land; it is also an occupation of the senses.[21] Throughout the book, then, I pay attention to how wind imaginaries forge certain soundscapes and touchscapes of Western Sahara.

Soundscape describes the sonic environment of our everyday lives.[22] *Touchscape* is less well theorized, but we might think of it as the daily sensations encountered by our skin. I am reluctant to refer to the tactile, because, although the wind imposes temperature and pressure on us, we cannot touch the wind. As Ingold puts it, "To feel the wind is not to make external, tactile contact with our surroundings but to mingle with them."[23]

What is the result of an absence of attention to wind imaginaries? Flikke argues that Westerners are unattuned to the multiple possibilities of the air, and the extent to which they are folded up with it. No one would survive long without breathing air, of course. It is part of us.[24] Flikke looks to Zulu Zionist ritual proceedings in South African churches to explore a different way of connecting with air and atmosphere. A focus on ways that different societies interact with the air offers new, vital possibilities for understanding climate change and the anthropocene, he argues.[25]

Researching the political role of moving airs is essential in this epoch of (climate) crisis. We are already seeing how some "green" developments continue oil's filthy legacy. From environmental damage and social marginalization surrounding the mining of rare earth materials for solar panels,[26] to livelihoods endangered by wind farms,[27] and dubious landgrabs to make way for biofuel plantations,[28] initiatives to tackle climate change often rely on the dispossession of, and violence against, already marginalized populations. With regard to North Africa, Hamza Hamouchene points out that the sorry saga of the Desertec Initiative is a fitting example of "green" extractivism.[29] The failed Desertec Industrial Initiative aimed to meet approximately 20 percent of Europe's energy needs by 2050 via solar and wind farms built in North Africa. Through such models, the production of wind energy moves to the periphery. As in the colonial era, the periphery bares the costs, while the Global North enjoys the fruits. Hamouchene highlights that local activists saw Desertec as a neocolonial capitalist endeavor. They raised concerns about the plunder of already scarce water resources, the export of energy to Europe without meeting local energy needs, and the colonial language developers used to describe the Sahara Desert.[30]

Focusing on wind infrastructural developments built on Indigenous lands in the Mexican Isthmus, Alexander Dunlap finds that so-called renewable energy would be better termed *Fossil Fuel+*. This is because the development of industrial scale "wind factories," as Dunlap and Martín Correa Arce term them, requires enormous amounts of mineral extraction and fossil fuel.[31] For Dunlap, "renewable energy," at least as it is developed by corporations, does not exist and is sustainable only in the sense that it allows the infinite accumulation and conversion of wind and sun into industrial and capitalist development along neocolonial lines. He sees the so-called renewable energy transition, as it is currently unfolding, as a war against the earth "being advanced by not only grabbing the land and ecosystems where turbines are placed, but the vitality of the air that they harness and convert into the circuits of capitalist infrastructure."[32] What, therefore, is to be done? Dunlap nods to the cultural. "If there is a 'solution,' it will depend on people's imaginations recreating their environments, reappropriating and transforming infrastructure to create, individually and collectively, healthy and sustainable environments."[33] He challenges all to "imagine alternatives" and suggests we might find inspiration in "indigenous knowledge."[34]

Imre Szeman and other authors of the Petrocultures Research Group, too, argue for our need to engage with Indigenous understandings of energy.[35] Scholars of this group, and scholars of the Energy Humanities more generally, have focused on oil's global permeation of culture. According to Sheena Wilson, Imre Szeman, and Adam Carlson, although oil is elusive and largely escapes cultural representation, all examples of cultural expression in the era of oil should be seen as crucially figured by this material since it underpins our political and social systems.[36] Similarly, for Brian Black, petroleum culture has "recalibrated global economies and political power" particularly over the last century.[37] For Cymene Howe and Dominic Boyer, focusing, like Dunlap, on the case of the Mexican Isthmus, corporate-led wind factories (to use Dunlap and Correa Arce's terminology) are developed with little transition attached to them.[38] They replicate the logics of a carbon economy that favors local and international elites at the expense of Indigenous peoples and local nonhumans. Likewise, David McDermott Hughes, in his study of wind energy developments in a southern Spanish village, finds that the infrastructure offers few benefits to villagers living among the turbines while further enriching wealthy landowners.[39] Graeme Macdonald argues that we can see energy structures as cultural objects. He shows the pipeline to be an object joining modes of high carbon living with politics, policy decisions, and energy infrastructure. How we think about, and act upon, the oil question is determined by how we see, narrate, and interpret oil, reasons Macdonald.[40] Similarly, focusing on water, Hannah Kate Boast argues that how we imagine water is inseparable from how we manage it.[41] At the time of writing, research on solar imaginaries is emerging. Rhys Williams has explored how solarpunk imaginaries offer inspiration for a better, fairer way to realize the energy transition,[42] while a special edition of *South Atlantic Quarterly* offers further initial academic investigations into solar cultures and imaginaries.[43] The first chapter of Howe's *Ecologics* ethnography shows how Zapotec cosmologies and multinational energy companies have very different conceptualizations and understandings of the wind, which shape the development of, and resistance to, wind developments in Oaxaca, Mexico.[44] However, there is as yet no literary or cultural studies research that explores how wind imaginaries, historical and current, shape energy systems in the ways that Macdonald and Boast respectively argue are true for oil and water imaginaries and developments. This work is crucial if we are to ensure a just energy transition.

Larry Lohmann offers a more radical argument. He asks if we really want the energy in "energy transition." He and others such as David Nye and Cara Daggett contend that today's dominant understanding of energy has emerged from nineteenth-century developments in thermodynamics, through which understandings of energy became organized around industrial, capitalist work.[45] Energy is white—it is born of white supremacy.[46] As an example, Lohmann takes the first law of thermodynamics, which theorized the steam engines that capital used to appropriate the maximum surplus from workers. The notion of energy that is hegemonic today was born of industrialized capitalism, globalized wage labor, and work as commodified labor-power, argues Lohman. He says, "From its birth, energy has been about a particular pattern of exploitation of human as well as nonhuman beings."[47] With the huge concentrations of power in fossil fuels globally, the measurement and commodification of this sense of energy could feed the illusion of infinite economic growth. Since fossil capitalism has defined what we mean by energy, Lohmann therefore argues that merely swapping fossils for another source of energy will not solve but will rather maintain the crisis of capitalism and the inequalities it breeds globally.[48] On this point, Lohmann takes issue with mainstream environmental organizations (and indeed even some more radical environmentalist movements). Lohmann points out that climate activists in the Global North too often expect those in the Global South to leave behind their "identity politics," accept that there is "no time" for questions of social justice, and get on with the most urgent of jobs: tackling the climate crisis. The problem of such Global North environmentalists, according to Lohmann, is that their conceptualization of climate is as white as their conceptualization of energy. People such as the Saharawis, suffering—as I show in chapter 4—at the hands of "renewable" developers, should forget their anti-colonial, anti-racist, and anti-sexist struggles and support the corporate wind and solar developments since the climate is more important than freedom or peace. The problem with this white climate activist understanding of the Saharawis' situation is that the climate crisis has been caused by the unrelenting colonialism, capitalism, classism, and patriarchy against which the Saharawi struggle is pitted. For Lohmann, the solution is a political transition, led by peasants and Indigenous peoples, that better recognizes plural forms of energy.[49]

Lohmann's arguments link to Meredith de Boom's conceptualization of climate necropolitics, through which violence is rendered legitimate in the name of climate change mitigation.[50] De Boom draws on Mbembe's necropower, "the power and capacity to dictate who may live and who must die."[51] Through necropolitics, a colonial sovereign power controls and makes disposable a "savage" population. If the latter must die or suffer such awful conditions as to constitute a living death to protect colonial interests, so be it. In climate necropolitics, as in Lohmann's conceptualization of white climate change, the communities that have least contributed to the climate crisis will suffer its consequences most harshly. Colonized peoples, Indigenous peoples, peoples like the Saharawis. These peoples are expected to pay the highest prices of the climate crisis, including violence, *as if such prices were inevitable*, in the service of civilization. One of the contentions of my book is that such prices are not inevitable.

In response to the predicament of neocolonial solar and wind factories, Diana-Abasi Ibanga advocates for recourse to an "African energy ethics," the purpose of which would be "to provide a rigorous Indigenous ethics for energy transitions in Africa."[52] His own contribution is to reflect on how energy infrastructure could be developed in a way that respects ethical and philosophical frameworks from various African communities, such as (to follow his own case studies) South African Braai ethics and Igbo *Ibuanyidanda*. Ibanga summarizes Braai ethic as follows: "A thing or action is right if it allows other existents to share one's space and resources; it is wrong if it tends otherwise. That is, the less a thing brings existents together in complementary sense the less right it is."[53] *Ibuanyidanda* is summarized by Innocent I. Asouzu as "a philosophy of integration and complementation."[54] For Ibanga, care between existents—human and others—is paramount for a just transition.

There is a consensus, then, among humanities and social science scholars interested in fairer ways to transition to renewable energy, firstly of the need to learn from the world's Indigenous populations, and secondly on the root of the problem: corporate-led solar and wind factories are replicating the capitalist, extractivist logics of fossil fuels. Linked to this, our very understanding of energy emerged from the industrial revolution and the colonialism it fueled. However, I argue in this book that there is another factor that has determined the historical trajectory of energy injustices: climate and weather imaginaries, and therefore sun and particularly wind imaginaries, were at the center of the colonial project from the outset, pre-industrial revolution. Lohmann rightly argues that the historical evolution of today's hegemonic concept of energy helps to explain fossil fuel and oil's bloody trajectories and that, therefore, a simple energy transition to renewables will do little to end energy injustices. I add that colonial understandings of wind, and histories of European and American interpretations of it (wind always has its own political history), also negatively influence wind infrastructure's revolutionary potential. Resisting wind infrastructure's injustices therefore must involve reimagining colonial conceptions of wind as well as of energy and understanding how historical wind imaginaries have shaped hegemonic conceptions of energy today.

Ahead of the age of steam, colonizers and traders relied on wind and sail for locomotion around the globe. The trade winds enabled and shaped European colonization, including the transatlantic slave trade. With reference to the case of Spain's Pacific empire, Greg Bankoff names the prevailing weather systems the "winds of colonization," since wind patterns determined which territories the Spanish colonized.[55] Any historical appreciation of Hispanic colonialism and culture without consideration of the role of wind would be incomplete, concludes Bankoff.[56] The effects of Pacific typhoons inland, argues Bankoff, also shaped the cultures of those who lived in those territories in terms of architecture, crops grown (an emphasis on root crop agriculture), and locations of settlements.[57] I argue that wind's role in shaping culture is stronger still, affecting the latter not only in terms of the physical manifestations of which foodstuffs are planted or where humans settle but also at the imaginary level. Turning to Europe's colonial adventures in North Africa, my contention is that a history of representations of Sahara winds can offer insights into wind's discursive role in (resisting) colonization and shaping colonial energy developments. Just as the works of energy humanities and social science scholars discussed above look to oil imaginaries and cultures in their analysis of unjust wind and solar developments, I am interested in what we can learn about fossil-fueled, as well as wind and solar, energy developments if we look to historical and cultural wind imaginaries. Wind shaped, and continues to shape, colonialism and its (energy) infrastructures in ways that go well beyond the nautical possibilities and perils of the trade westerlies and easterlies in the Age of the Sail.

Briefly then, the thesis of this book is that wind imaginaries have been at the center of Europe's colonialism from the Age of Sail until today, with tangible consequences for energy infrastructure and use. A fairer future, in terms of renewable energy, therefore requires a departure from colonial wind imaginaries. In this book, I aim not only to show precisely how changing wind imaginaries shaped (and were shaped by) colonial (energy) projects in Western Sahara, but also to explore alternative Indigenous Saharawi wind imaginaries, what role these have played in resisting colonialism, and how these may positively shape Western Sahara's energy future.

My methodology is necessarily interdisciplinary. It evolved over the course of the project in line with changes in personal (the birth of my two children), regional (the end of the cease-fire in Western Sahara), and global (the COVID-19 pandemic) circumstances. A field trip to occupied Western Sahara became impossible. However, I do partially rely, in part 2 especially, on extensive notes from earlier (2014) field trips to occupied Western Sahara, when the idea for this project was conceived. During these earlier trips, undertaken to research and document gendered experiences of occupation, resistance, and human rights abuses, I observed how my host family adapted their everyday lives to cope with power and water cuts. I asked interviewees about their perceptions of the energy system and former nomads about the local winds. I began to understand the strategic use of power cuts as political tools. Nevertheless, given the absence of a recent field trip, chapter 4 also relies on interviews and focus groups carried out by my colleagues Hamza Lakhal and Mahmoud Lemaadel. This chapter is an updated and revised version of an article that the three of us co-researched, and which I wrote up on our behalf, included here with their permission. The ethical implications of research in occupied Western Sahara are substantial due to the risks assumed by participants. My conviction is that there is an ethical argument to carry out research in occupied Western Sahara *as long as* the researcher (or activist-scholar) allies themself with the oppressed Indigenous population in a meaningful, committed, long-term way. The ethical discussion is as complex as it is important. It is too long to include here. Instead, I refer readers to an earlier publication on this issue.[58]

My methodology also evolved in response to my misconceptions about the size of the current energy system in Western Sahara. I always planned to use cultural studies methods, such as literary analysis, applied to colonial and Indigenous texts on Saharan winds. But I needed to combine this with methodological tools that could also speak of energy system infrastructure and management. I quickly discounted an ethnography of the Saharan winds. Although such a project appeals to me immensely, surely only a Saharawi nomad would have the knowledge to do such a project justice. Instead, I hoped to follow the energy. I wished to carry out an ethnography that would take me from the occupied territory, following the submarine energy cables that join the Moroccan grid to the European one via Tarifa, Spain and then going southbound to the Mauritanian grid. However, a month into the research I found that, if I were truly to follow the energy, I would need to travel to an infinite number of places across the globe, which would be unmanageable. Turbines from the Basque country, LPG from North Sea oil fields, a potential pipeline to Nigeria—powering the Moroccan occupation of Western Sahara is a transnational feat. Therefore, although I nod to the transnational hubs that power the energy system in Western Sahara (the opening vignette at Teesside port is one such nod), my research homes in on the Saharawi Arab Democratic Republic. This includes five research fields defined by space and time: 1) the territory within the official UN borders of Western Sahara, which was ruled by no one and roamed by "wandering Arabs" during the Age of Sail, 2) "Spanish Sahara," the territory colonized by Spain in 1884, 3) modern-day Moroccan-occupied Western Sahara, 4) the modern-day "liberated" Saharawi Arab Democratic Republic or "free zone," and 5) modern-day SADR-in-exile.

Research across these five fields has relied to varying extents on cultural studies methods of close reading and textual analysis of poetry, other literary texts including novels and memoirs, and nonfiction science and pseudoscience writings. For chapter 2, I carried out archival research at the Spanish General Administration Archive, which houses government documents from what was Spanish Sahara. Research for chapter 8 is based on a field trip (2019) to the SADR-in-exile. I used the trip to research existing and planned energy infrastructure in the refugee camps/SADR-in-exile and the free zone and also to learn more from nomads about the local winds. I also make use of notes from previous trips (2006, 2008, 2015) to the camps and free zone. Occasionally, I draw on my experiences of volunteering for Western Sahara Campaign UK, Western Sahara Resource Watch, and Polisario's representatives to the UK and EU. Respectively, my work with each has focused on human rights, lobbying against legally questionable natural resource exploitation activities, and assisting in the development of the SADR's indicative Nationally Determined Contribution (iNDC) to the climate agreements.

This book has four parts. Part 1 looks back to the lows of European colonialism: the Age of Sail (chapter 1) and Spain's annexation of Spanish Sahara (chapter 2). If we are to understand the contemporary climate crisis, we need to understand its historical context: how climates have been measured, imagined, and used in the past, for imperial projects. Likewise, if, as Ingold argues, air is a "material medium that has escaped the bounds of materiality," then how did wind come to be perceived as a resource? What sort of resource-making project achieved this?[59] The oldest use of solar power is agriculture—and of wind power, the sail. Chapter 1 turns to eighteenth- and nineteenth-century written testimonies of European and North American sailors shipwrecked on Western Sahara. The history of European colonialism is a maritime history. It is therefore also a history shaped by wind. The coast of the territory today called *Western Sahara* was once a nursery for Portuguese sailors learning to seek out the so-called trade winds in order to travel in the desired direction. The same coastline also features Cape Boujdour, the place that marked the end of the world for medieval European navigators. We do not know exactly how many European and North American ships the winds and currents destroyed at the cape: to report a wreckage, there had to be at least one survivor. Several such survivors wrote detailed, book-length accounts of the circumstances of their shipwreck and their subsequent capture by Cape Boujdour locals, whom they called the *wandering Arabs*. Such books were often bestsellers when they were first published and still influence popular culture today. For example, Russell Crowe was recently set to star in a Hollywood film *In Sand and Blood*, based on the bestselling nonfiction book *Skeletons on the Zahara: A True Story of Survival* by Dean King.[60] King's book, in turn, was based on North American sea captain James Riley's 1817 memoir *Sufferings in Africa*, which charts the wrecking of his ship *The Commerce* in what is today known as Western Sahara and his time living as a captive of Arab nomads, the ascendents of people we today call *Saharawis*. The book was an international bestseller in its time and was listed by Abraham Lincoln as one of the three books that most influenced his political ideology. Memoirs like Riley's shaped contemporary, popular understandings of the Sahara Desert and its people. And, I argue, both the desert and its people were imagined as windblown. In chapter 1, I analyze testimonies of European- and North American-penned memoirs of shipwreck and captivity in Western Sahara in terms of the meanings they attach to the Saharan winds, and how these informed and drew on wider discourses that attempted to justify European colonialism.

When Spaniards arrived in Spanish Sahara from the late nineteenth century and into the twentieth century, they looked down. They wondered about what lay below—phosphates, oil, things of no consequence to nomads at that time. Saharawi nomads looked up and across as well as down. They did so in order to navigate. For nomads, celestial orientation intermixes with the topological: constellations are mapped on geological features, themselves the result of the constant intermingling of winds and lands. Epistemological wind politics, or more specifically, Saharawis' knowledge of the desert and its winds on the one hand, and Spanish absolute dependence on Saharawi nomads due to their own lack of knowledge on the other, is at the heart of chapter 2. I use Spanish colonial government archives, research publications, and other nonfiction writings as my sources. For most Spanish colonizers, and for the sailors before them, the strong and hot Saharan winds, sometimes called the *red winds*, were a metaphor for all that is barbaric. In Western Sahara's case, wind and aeolian geomorphology played a primary role in justifying colonialism. Because the winds and aeolian geomorphology were seen as terrifying, pathological, and deadly, Spanish infrastructural developments were viewed as a colonial triumph over hostile nature. I argue that negative imaginaries of the wind and of the windblown—aeolian anxieties—underpinned (energy) infrastructure in Spanish Sahara and indeed underpinned the entire Spanish colonial project. The political regime mediated by this energy infrastructure reproduced racial differentiation and exclusion of Saharawis. As the Saharawi nationalist, pro-independence movement emerged, the colonial energy system became one more site for expressing anti-colonial dissent.

In part 2, I turn to modern-day occupied Western Sahara and the home of my friend Ali's[61] extended family. It was when visiting Ali that I first came to think of power outages in Western Sahara as a state-orchestrated politicized aggression. But one cannot just visit a Saharawi friend in occupied Western Sahara as you would a friend in most other countries. To make it to a friend's house without first being removed by the Moroccan state is a challenge—a challenge that spotlights the political situation and, I found, the place of (renewable) energy infrastructure in it. The process of attempting to travel to occupied Western Sahara therefore becomes part of my narrative in chapter 3. The chapter's focus is the wind imaginaries of the Moroccan-occupier's corporate energy partners in modern-occupied Western Sahara. Using Public Relations (PR) materials published by Siemens and Siemens Gamesa between 2009 and 2017, I argue that modern corporate imaginations of Saharan winds replicate typical European colonial discourse on colonized lands and people. Corporations imagine the Saharan winds as wild, barbaric, backward, and in need of colonial tutelage, modernization, and development. Energy developers *tame* the winds. The differential problem that wind factories bring for colonized peoples is their tremendous PR potential and veneer of legitimacy. As Tom White points out, renewable energy comes with an inbuilt "ethical cache," in which the rhetoric of environmentalism is mobilized to serve "green" capitalism.[62] While oil shyly hides beneath the earth's surface, offshore or in underground pipelines, and oil barons work hard to hide their environmental crimes, wind and solar corporations lose all modesty in showy pictures of their mills and panels, saturating their PR campaigns with images and videos of farms and developments, even lending their blades for art installations.[63] When it comes to the green capitalism and colonialism of energy corporations, the battlefield is therefore that of discourse, meanings, cultural representations, and understandings.

The *Tallīya* is a northerly wind. If you are a Saharawi fisherman, it is a good omen. There will be fish. If you are a Saharawi nomad, the coming of the *Tallīya* means it is time to dig trenches to avoid your *ḫaīma* flooding. The significance of the *Tallīya* is a matter of perspective. In chapter 4, I discuss Saharawi perspectives of the (wind) energy system in occupied Western Sahara. I use field notes and interview transcripts from my 2014 field trip there. I also use more recent interview transcripts and focus groups carried out by Hamza and (as mentioned above) another colleague of ours, Mahmoud, which concern Saharawis' perceptions of the energy system in occupied Western Sahara. The chapter focuses on the politics mediated by the energy system in occupied Western Sahara, an energy politic that is entangled with corporate imaginaries of Saharan winds. I argue that colonial wind imaginaries breed colonial energy politics. To illuminate the violence and racism of the energy system, I follow the stories of my 2014 hostess, Ali's relative Safia, who must cope with the trials of bringing up children without reliable electricity and under the constant threat of political oppression; Sultana, who leads an organization that protests the energy developments and has had her eye removed by police as punishment for her activism; Zrug, who has worked for both foreign hydrocarbon and wind companies active in Western Sahara but who has become radicalized against such developments; and others.

Part 3 refers, like part 2, to modern times. But I move away from colonial corporate imaginations of Saharan winds and look instead to Saharawi perceptions of the latter. I situate myself in the beloved Saharawi desert heartlands: the *bādīa*. My main source for part 3 is poetry, complemented by interviews with Saharawi nomads and writers. Chapter 5 makes use of the Spanish-language poetry of the Friendship Generation of Saharawi writers, who have lived most of their lives in exile from Western Sahara, including in Cuba for their adolescence and young adulthood. The winds, in Friendship Generation poetry, recall the beauty of a faraway, longed-for, and loved place: they plunge down gilded dunes, whoosh through dried-up wadis, and reach into the shadowed places of treasured *gallāba*. But as well as sometimes constituting the subject and focus of poetry, I argue that the Saharan winds also inspire the aesthetic tools of the poets and that the poets' wind imaginaries challenge the corporate colonial discourses explored in previous chapters. The poetry also offers the reader, I contend, glimpses at Saharawi epistemologies by giving insights into how nomads read the winds, and wider aeolian geomorphology and processes, which allows their survival in the desert. In particular, I use poetry by, and interviews with, Limam Boisha and Fatma Galia Mohammed Salem to illuminate how Saharawis use wind and aeolian geomorphology to navigate the desert and learn of imminent threats beyond the next dune.

The *īrīfi* is an easterly wind, which, for Spanish colonizers and European and American shipwreck survivors (the latter knew it as the *Shume*), blew with terror-inducing, fierce heat. *Īrīfi* is a *Ḥassānīya* word, which is probably borrowed from the similar Berber for hot wind, extreme thirst, or to suffer from the heat.[64] Of course, Saharawis too are vulnerable to the scorching sandstorms of the *īrīfī*, but nomads know how to foresee, and prepare for, the wind's arrival in order to minimize risk. Nomads can read the desert and its winds. Or, perhaps rather than read, they listen to what the desert more-than-humans advise. This chapter focuses on nomads' epistemologies of Saharan winds and other desert nonhumans, and on the place of Saharawis in the desert ecology, using poetry as my main source. I focus on poets and artists that, unlike the Friendship Generation, spent their formative years in Western Sahara. This includes both younger writers born in the occupied territory and older nomad-poets (writing or composing in Arabic or *Ḥassānīya*) born in Spanish Sahara, as well as the work of an occupied Western Sahara-born filmmaker who draws heavily on the work of nomad-poets. In chapter 6, I build on the previous chapter, further exploring the use of aeolian aesthetics and the use of wind as a metaphorical resource for Saharawi poets.

Part 3 focuses on Saharawi imaginations of Saharan winds, and part 4 asks what such imaginations might mean for the future of energy. Saharawi poets and artists challenge the colonial and capitalist conceptions—promoted rigorously by Siemens—of energy, sun, wind, and desert. But as they do so, do they also take on the common challenge at the root of all energy injustices, no matter the fuel, which is to reconceptualize energy? The *Gblīya* is among the most popular winds in the camps. It is a harbinger of good news, a sign that light rain is on its way and a metaphor for all that is fair and hopeful. In chapter 7, I explore the extent to which Saharawi wind imaginaries might promise a more just energy future. I focus on emerging and planned energy systems in the state-in-exile in Algeria. The chapter relies on ethnographic research carried out in the Saharawi state-in-exile, exploring the tensions between the long-term use of precarious solar panels versus recent connections, in some camps, to the Algerian electricity grid. I also focus on—via interviews with electrical engineers, policymakers, and politicians—the Saharawi Republic's energy policies and plans for the liberated territory.

In February 2021, European media outlets reported on the bothersome Sirocco wind blowing in from North Africa and carrying sand and dust. The latter provoked motorway speed restrictions, pollution warnings, and concerns over radioactive particles as far north as France. This media focus on the physical encroachment of the Sahara Desert, via its winds, in Europe arguably stands in for the wider European hysteria of the intrusion of illegal immigrants from the Sahara and beyond. Mainstream European media meanings attached to Saharan winds are anxiety-driven and hostile. But Saharawi refugees living in Spain and their Spanish allies use the mobile possibilities of wind to imagine a rather more positive encounter between Europe and the Sahara. Chapter 8 focuses on such encounters. I discuss Spanish and multiauthored Spanish/Saharawi novels, stories, and material art produced since Spain's exit from Western Sahara in 1975 that engage with the former colony. I explore the meanings attached to Saharan winds in the works, especially in terms of how they are used to evoke solidarity with the Saharawi population, and ask whether transnational solidarity might underpin a fairer energy future.

Back in Teesside, a building breeze encourages us to go home. I make a note to write to the port authority once we know for sure that the vessel arrives in El Aaiún. From the main road, we see seals hauling themselves onto the Teesside mudflats and a peregrine falcon perched on top of a nuclear reactor. I wonder what my oldest boy thinks the wind is. "A ghost," he says, and then asks when the boat we have spotted will reach "Uncle Hamza" in Western Sahara. Out to sea, the turbines of the Redcar wind factory whirl.

Part One

The Trade Winds

CHAPTER 1

The Sea without Water

Navigators, Traders, and Wind Pathologies in Western Sahara

Nothing but calm, green waters pawing at the shore behind him, Gil Eannes of Lagos stood in his wet boots, whitened with sea salt, and surveyed the sandy horizon for signs of life. He recalled the reception at court—as stormy as the weather—upon returning from the previous year's voyage. Last time, the winds betrayed him. They forced his boat to one of the Canaries, northwest of his desired destination, and not even the capture of a few Indigenous islanders could placate his sponsor, the Portuguese prince.[1] The Infant Henry had freshly promised Eannes glory and wealth should this current expedition be successful. "Make the voyage from which, by grace of God, you cannot fail to derive honor and profit," the pious Prince, in his hair shirt, had ordered.[2] Now, Eannes had completed his mission to pass Cape Boujdour, a feat widely considered impossible and one that expanded the European navigable world forever. But how to prove it to the Infant?

The moist coast mists resurrected the tangled-up, brain-shaped moss. It unfurled its curling fronds, which turned from a dead brown to a radiant green—ends blushing sunset pink—and softened in texture.[3] St Mary's Rose, the most resilient of plants, thrived in deserts. Able to curl into a dry ball to protect itself when conditions were unfavorable and bloom out again when rain came, and possessing no need to anchor roots, the moss would survive the journey back to Sagres and prove that its carrier had trod the desert beyond Boujdour. Eannes scooped up the specimen and packed it carefully into a small wooden barrel.

For centuries, Cape Boujdour marked the geo-epistemological border of Western knowledge of Africa. Medieval Europeans saw the headland as the end of the world, the place where ships and their crews disappeared forever. Its name deriving from Abu Khatar, Arabic for *father of danger*, Cape Boujdour was a source of fear, ferocious legends, and speculation for sailors. Inland, the Sahara Desert was just as hostile as the dangerous cape, impassable to all. Fifteenth-century mariners, in the age when Europeans tried in earnest to pass the promontory, believed that "beyond this Cape there is no race of men nor place of inhabitants: nor is the land less sandy than the deserts of Libya, where there is no water, no tree, no green herb- and the sea so shallow that a whole league from land is only a fathom deep, while the currents are so terrible that no ship having once passed the Cape, will ever be able to return."[4]

The land of modern-day Western Sahara and its coastline were understood together as one great no-man's-land. Ghislaine Lydon argues that the long-term consequence of this European-imagined great barrier is the flawed division of North Africa and sub-Saharan Africa into diametrically, culturally, and racially opposed Africans, between which the supposed border—the Sahara—has become an understudied blind spot.[5] But in the Middle Ages, this part of the world was infamous, potentially promising a sea route to the East Indies, and, on land, access to a legendary "River of Gold."[6]

It was not until 1434 that Gil Eannes, a squire of Portuguese prince Henry the Navigator, rounded the terrible Cape Boujdour and made it back to tell the tale.[7] Chronicler Gomes Eannes de Azurara recorded this feat and other Portuguese exploits in "Guinea."[8] *Guinea* was the collective name for the lands south of Morocco, which was unknown to Europeans up to this time. Azurara's texts were the first to focus on the experiences of Europeans in Africa south of Cape Boujdour. These texts also documented the start of the transatlantic slave trade.[9] David Hughes has described this abhorrent trade as "the first intercontinental energy market."[10] Hughes argues that Spanish enslavers in Trinidad "invented fuel" by way of exploiting human labor on the island's plantations, since fuel is energy that is stored in a measurable, transportable, and salable form.[11] Later, charts Hughes, eighteenth-century Spanish colonists in Trinidad proposed a "solar colony" in which cacao, pulled up by the sun, would enrich Spain.[12] He shows that solar energy and fuel more widely have a history tangled up with colonialism and enslavement.

But the history of European colonialism is also tied up with the wind. Seafarers of the Age of Sail had to learn which winds and currents could take them to their desired destination. In the fifteenth century, the so-called

Atlantic Mediterranean, roughly comprising the seas, winds, and currents of the coasts of Morocco, Western Sahara, the Canary Islands, and the Azores, became a nursery for would-be imperialists.[13] Off Western Sahara's coast, Portuguese sailors learned to seek out the trade winds in order to travel in the desired direction.[14] This movement, known as the *Volta do mar*, was a vital step in the history of navigation and enabled further European exploration of the Atlantic and Pacific Oceans and the resulting imperial pursuits.

It was also the terrible northwest African winds and currents that forced European ships onto reefs and made Cape Boujdour impassable for so long. This cape was a hotspot for wrecks throughout the Age of Sail. Several Portuguese navigators that attempted to follow Eannes's route soon after him were unable to land due to "contrary winds" and "tempests."[15] For the Portuguese that made it into the desert, the inland weather was just as worthy of chronicle. Azurara noted, for example, the fascinating findings of Joao Fernandes. Fernandes was a curious fellow among his countrymen in that he was uninterested in capturing booty and enslaving Africans but urged his prince to be left in the Sahara for several months, where he hoped to learn the ways of the Indigenous peoples. An Arabic speaker, he lived for seven months with the inland nomads, where "the heats of [the] land [were] very great,"[16] and was amazed that "those Moors with whom he travelled guided themselves by the winds alone, as is done on the sea, and by . . . birds."[17]

While Fernandes marveled at Indigenous uses of wind inland, his compatriots, under instruction from the Infant Henry, meticulously recorded weather patterns off Cape Boujdour and beyond. Consequently, Portugal developed a priceless inventory of northwest African winds, tides, and currents, which it was notoriously reluctant to share with other European powers.[18] Wind was the key to the imperial race.

Despite medieval explorers' insistence in documenting African weathers, Sverker Sörlin and Melissa Lane find that most humanists and social scientists have not yet found it relevant to engage with climate issues.[19] This is a large omission. A focus on how historical subjects understood, imagined, used, and talked about aspects of the climate—in Western Sahara's case, the wind above all and to a lesser extent the sun—is integral, I argue, to understanding the colonial project in general and the colonial hegemonic constructions of wind energy more specifically. In the next two chapters, therefore, I focus on representations of solar and especially wind power, first, in the chronicles of the exploits and written testimonies of European navigators, enslavers, and merchants who found themselves in Western Sahara, and second, in the written works of Spanish colonialists of the Sahara. I argue that wind, sun, and the sand they blew and heated were key to the white construction of the colonial *other*, informed colonial projects in general terms, and, when the Spanish colonizers built energy infrastructure, wind imaginaries shaped the latter's development and management. Allying themselves with the Indigenous Saharawis, wind, sun, and sand also undermined colonial discourse from within.

Wrecked on Cape Boujdour

James Grey Jackson was a British merchant who lived in Morocco for sixteen years. He poured the knowledge he acquired there into his major work *An Account of the Empire of Morocco and the Districts of Suse and Tafilel*. Writing in 1809 of the period 1790–1806, and warning his readers of the perils of the coastline south of Morocco, he noted that there were thirty wrecks on Cape Boujdour from which at least one crew member survived to tell the tale. Of these, seventeen were British, five French, five American, one Dutch, one Danish, and one Swedish.[20] Jackson also suggested that these thirty wrecks were anomalies in that they were recorded at all. Likely, most wrecks went unreported—their only witnesses the perished passengers.

The written testimonies of those who survived shipwreck off Cape Boujdour fit squarely into two genres at once: the shipwreck narrative and the captivity narrative. The infamous and incredibly popular captivity narratives of white North Americans and Europeans captured by Indigenous North American Indians were preceded by the lesser known (but, in the seventeenth and eighteenth centuries, immensely popular) Barbary narratives of Europeans and Americans captured by North Africans and sold for ransom. For example, Captain James Riley's account of shipwreck at Cape Boujdour followed by his captivity was a bestseller, came out in at least twenty-eight editions including an illustrated children's edition, is still in print today, and continues to inspire literary and film spin-offs.[21] Real-life shipwreck accounts were, like captivity narratives, hugely popular. The first of this genre appeared in Portugal in the sixteenth century, and, with the expansion of print culture throughout Europe, grew steadily in popularity in the major maritime nations throughout the seventeenth, eighteenth, and nineteenth centuries.[22]

Josiah Blackmore argues that shipwreck literature undermined hegemonic historiographic narratives of imperialism. The shipwreck narrative is unsettling and disruptive because it "promotes breaches in the expansionist mentality and in the textual culture associated with that mentality."[23] From the time of Columbus well into the nineteenth century, the ship was the vehicle of Western culture, technological superiority, and political might, and it was the medium through which the West encountered and eventually domesticated "barbaric" and "uncivilized" communities.[24] As Carl Thompson points out, Western cultures were maritime ones, and ships played a central role in generating the "scientific" knowledge on which these cultures prided themselves.[25] Therefore, testimonies of the wrecking of European ships in the age of European colonialism provided a recurrent reminder of the limits of imperial power. Wrecks exposed ruptures in Western global trade networks and attempts to develop knowledge and power.[26] But what caused the shipwrecks and the resultant undermining of Western imperial discourse? If ships were the principal vehicle of imperialism in the Age of Sail, then weather—above all, wind—was not only its fuel but also, in the case of wrecks, a reminder of imperialism's limits.

Mariners identify approaching winds via dancing waves, scraped up from the sea's surface by the winds. They nickname these wavelets *cat's paws*.[27] Around Cape Boujdour, the paws shift direction. They begin to crawl from the northeast. Meanwhile, desert winds carry loose sand from immense dunes inland—themselves created by winds—for miles out to sea, creating a misleading haze. In the Age of Sail, the strange mesh of airborne sand left navigators unaware of how near land they were "until they discover[ed] breakers on the coast."[28] Added to this, explained Jackson, "there [was] a current, which sets in from the west towards Africa, with inconceivable force and rapidity," not to mention shallow waters reaching far beyond the coast.[29] Charles Cochelet, wrecked on the *Sophia* in 1819 upon his captain becoming "baffled by the winds," warned of a "fatal" current leading toward the African shore from the southernmost point of Fuerteventura and of an "atmosphere . . . so impregnated with the loose sand of the Sahara, impelled hither by the wind, as to obstruct the view of objects at a distance."[30] Likewise, Captain Riley of the wrecked North American brig *Commerce* wrote, "The thick foggy weather would prevent our seeing the land in the day time: whilst the wind, blowing almost directly on the land, would force us towards it, and endanger the safety of both the boat and our lives at every turn or point."[31]

As Christine Sears points out, shipwreck and captivity survivors' narratives were tailored to sell to a public and—frustratingly for Sears—contained several personal and cultural biases.[32] These cultural biases are themselves of interest to me. They mean that, as well as thrilling spectacles, the published tales of white shipwreck and captivity survivors became another means for barbarizing Africans and therefore for justifying the continent's colonization. And in parallel, the written testimonies provided several layers of knowledge that further enabled colonization. The narratives of shipwreck and captivity on

Cape Boujdour that I analyze here all contain these double elements of simultaneously justifying and enabling colonialism. They do so largely through an orientalist framework. Orientalist discourses create a totalizing description of Islamic lands and societies based on negative descriptions of local communities and their social habits, governance structures, religion, and so on. The desired effect is to show the social superiority of the (Christian) West.[33] A second, equally important facet of orientalism is to suggest the West's potential mastery of native cultures. This is typically achieved by illustrating the West's alleged ability to *know* Indigenous cultures.

Henry the Navigator sponsored several successive missions to round Cape Boujdour in pursuit of knowledge—general knowledge of what lay beyond, knowledge of whether any Christians lived there, knowledge of the power of the infidels that might dwell there. Another aim of Henry the Navigator was to spread Christianity, with, he believed, God's support.[34] Centuries later, the European and American shipwrecked mariners took advantage of their stays to gather knowledge, which they transmitted in their narratives. Archibald Robbins, a crew member of the North American brig *Commerce*, for example, justifies the existence and usefulness of his own memoir on the basis of the innovative knowledge it offers. At the time of the publication of his text (1818), Robbins said that there were no writings on "the wandering Arabs" or on "the immense desert of the Zahara."[35] He found that European and American slaves (Robbins, like most mariners whose work I analyze in this chapter, questionably regarded himself as a survivor of slavery—even though he was not forced to work—and would have most likely perished had his "captors" not transported him to his nation's consular services in Morocco) were in a unique position to provide "correct information" on the desert and its peoples. Mere European or American travelers could not be trusted, argued Robbins, because they would be overly preoccupied with avoiding death or slavery. Slaves, however, had the "leisure [to] record in safety the peculiarities of this peculiar people" since their masters kept them safe.[36]

French merchant Monsieur Saugnier, wrecked off Cape Boujdour in 1784, is equally careful to justify his expertise on the lands of the "wandering Arabs." In the preface to his book-length account, which he directs to the French government, he suggests that the latter might commission him to take exploratory journeys into the interior countries of the Sahara, not yet trod by Europeans. To persuade his audience, he lists the seven principal barriers to a successful voyage and then explains how he is uniquely equipped to overcome each one. First is "the unhealthiness of the climate," which—he explains—his testimony of surviving slavery in the desert proves he is hardy enough to bear.[37]

Captain Riley, of the wrecked *Commerce*, ends his book-long testimony with an appendix on "observations of the winds, currents, &c., in some parts of the Atlantic coast of Africa," among other appendixes on Arabic vocabulary (including four different expressions for types of wind) and some letters.[38] These sailor-writers taxonomize: they attempt to contain Western Sahara and its people through knowledge. Captain Judah Paddock, shipwrecked on the Oswego in 1800, weaves such observations into his wider narrative. In chapter 7 of his account, for example, he describes the Saharawi manners of smoking, storytelling, marriage, hospitality, educating boys, prayer, appetite, and horsemanship.[39] French diplomat Pierre de Brisson was wrecked in 1785 on the *St Catherine* at Cape Blanco (now known as Ras Nouadhibou) in lavender-scented clothes and with soft-skinned hands. He reserves a special section in his book to describe "the religion, manners and customs of . . . the Arabs of the desert," with whom he spent fourteen months, detailing their marriage and funeral practices, how children are educated, the division of labor, the tent and its contents, and the Saharawis' physical appearance.[40]

A discussion on mariners' taxonomies of Western Sahara would be incomplete without a mention of their names for the Indigenous people themselves. All the mariners were careful to differentiate what they called the "wandering Arabs," "wild Arabs," "independent Moors," or "Mongearts" who take them captive from the trading Arabs to the north, who were subject to the rule and laws of the Moroccan emperor. Hector Black, wrecked on the Glasgow brig *Surprise* in 1815, wrote, "these lawless tribes of wandering Arabs [are] far from being in any degree under the government of the Emperor of Morocco."[41] Saugnier additionally differentiates the "independent Arabs of the Zaara" from the Arabs "subject to the dominion of the Emperor of Morocco" by joyfulness.[42] He finds those living under the draconian whims of the emperor "less happy" than the "desert wanderers."[43] Cock clarifies that the wandering Aboussebah (or "Sba") "present a barrier to the extension of the Emperor of Morocco's dominion."[44] To use the term *Saharawi* for these wandering tribes is an anachronism. Nevertheless, the term allows us to avoid repeating the orientalist labels of the European and American shipwreck survivors while maintaining the latter's observations of a distinct (even if the distinction was, like all category borders, socially constructed) people, ancestors of today's Saharawis. I follow other modern-day writers working on these texts in using the term *Saharawi* to describe the captors of the shipwrecked mariners.[45]

In orientalist texts, the ability to accumulate knowledge was and is evidenced by reverting to authoritative, "factual" discursive strategies in descriptions of the "natives" and their surroundings and by dividing and categorizing aspects of the Orient into manageable parts such as marriage customs, geography, and so on.[46] I argue in the following section that in European and American memoirs of travel, shipwreck, and captivity in Western Sahara, weather—principally wind—is central to both of these orientalist discursive strategies, denigrating the local population and land in ways that further colonialism but also undermine it.

Winds at Sea and Saharan Winds

In early May 1815, Captain James Riley's *Commerce* sailed from Connecticut bound for Africa. The 220-ton brig had paused at New Orleans and Gibraltar on the way, where the crew left their small haul of flour and tobacco and picked up brandy and wine. By late August, they were at sea again, headed for the Cape Verde islands, where they would acquire salt to trade back in Connecticut. Bound southwest from Gibraltar, the captain planned to sail through the small corridor between Tenerife and La Palma, but as they reached the archipelago, the Atlantic engulfed the brig in a thick fog. Night fell. Captain Riley had been unable to make his observations and calculate his latitude. When the sun returned and the clouds finally parted, there was no sight of Tenerife's twelve-thousand-foot volcano. The roar of breakers suggested that the brig had drifted east of the Canaries and that the currents had driven the vessel toward the North African coast. By the time Riley realized this, it was already too late: they were on the rocks. Hunted by menacing weather, the crew scrambled and scurried about the deck and, on Riley's orders, up and down from the hold carrying provisions. They hauled the latter onto the longboat and rowed for their lives to the nearby beach.

After the initial wreck, Riley and his crew, imbibed by myths of Indigenous cannibalism, decided to flee by longboat. They hoped to hail a passing ship at sea or land farther south, perhaps at the British-held settlement at Saint Louis, Senegal. In his memoir, Riley recalls the journey partially in the present tense. The sea swells lift and drop the longboat, but the crew's luck seems initially to take a turn for the better: "The wind now veered four points to the eastward, so that we were enabled to fetch past the point of the Cape; though the boat had neither keel nor rudder, it was sunset when we got out, and night coming on, the wind as usual increased to a gale before morning, and we kept the boat to the wind by the help of an oar."[47]

The wind around Cape Barbas (south of Cape Boujdour in Western Sahara) is described with empirical detail in terms of its direction, while Riley's apparent ability to navigate the wind even in difficult circumstances reveals his skill and knowledge. There is a similar dynamic in other testimonies by captains, but crucially only during the episodes at sea. Captain Paddock, of the *Oswego* (wrecked in 1800), for example, makes detailed observations of the winds throughout the first chapter of his memoir, the only chapter in which most action is at sea. Paddock writes of the wind as he would in his logbook, noting its direction in abbreviated terms—"N.N.E.," "wind varied suddenly to N.N.W."—and with nautical vernacular typical of the day, giving the wind character: "fair," "fine," "damp," "fresh."[48] Sailors' logbooks recorded log-derived speed, compass-given direction, wind direction, and, often commenting on weather, "remarkable occurrences." This logged information aided sailors' sense of the ship's course and location.[49] As Sorenson points out, eighteenth-century sailors' written accounts of voyages were informed by the language and content of logbooks,[50] and accounts of shipwreck in Western Sahara were no exception. Including technicalities of wind direction and force helped to bestow authenticity on mariners' narratives.[51] These precise climate details also reveal the discursive strategies of empire, which rely on voices implying factual or empirical authority.[52]

Returning to Riley's narrative, at moments when he is unable to navigate or use the wind, the air current becomes a character, and grammatically an agent, in his narrative. As his hopes for escape by longboat diminish, Riley writes, "The thick foggy weather would prevent our seeing the land in the daytime: whilst the wind, blowing almost directly on the land, would force us towards it, and endanger the safety of both the boat and our lives at every turn or point."[53] By making the wind a decided agent, Riley excuses himself for not having been able to successfully navigate the element. Yet, in making the wind an agentive character, he maintains his ostensibly deep knowledge of the wind. He is still able to describe (and therefore know) the sea wind in detail, even if he cannot control it.

Captain James Irving is likewise careful to describe the winds at Cape Boujdour with mathematical precision. Irving, of Langholm, Scotland, had his first captaincy on board the maiden voyage of the *Anna*, which sailed from Liverpool on May 3, 1789, bound for the Gold Coast. Captain Irving was an enslaver. The *Anna* carried a hoard of hardware goods, salt, and one thousand dollars with which to purchase enslaved peoples. His journal entry for the day of the *Anna*'s wreck contains several logbook descriptions of the wind—"a light breeze from the S.E.," "the wind which blew fresh at North or N.B.W." It also uses a plethora of verbs in the passive voice, which implies an agentive and manipulative local weather that plots against the captain and his men. Irvine is "deceived" and "persuaded" by the currents.[54] The victimhood suggested by the passive voice combined with the logbook-style accuracy in descriptions of the winds serves to excuse Irvine of culpability. As Suzanne Schwartz, editor of Irvine's journals and letters, suggests, his account is potentially crafted to avoid accusations of poor seamanship.[55]

Paddock, too, gives the wind agency at sea. He sums up the nautical art of navigation as weighing up one's "opinion as to the heave of the sea and the ship's craving the wind."[56] At the point of wreck, the wind is the protagonist of the narrative, bewitching the ship to cause the crew's downfall: "The main or mizzen-top-sail kept shivering or edging to the wind. . . . When she began to gather head-way, the helm righted with the wind at least two points on the starboard quarter, wanting not more than once her length of coming round, heading off shore."[57] On the beach Paddock, like Riley, hopes to escape by sea. But, also like Riley, his plans thwarted, he and his crew are captured by "wild Arabs."[58]

For the Saharawis, looting wrecks and capturing surviving sailors was a lucrative business. Westerners could be taken for ransom at a European consulate in Mogador (modern-day Essaouira in Morocco) or sold on to other Arabs, or citizens of the great African emporium of Timbuktoo. Thus, unlike Europeans captured in Algiers who were used for their labor, those taken in Western Sahara were rarely put to work—they were valuable for their potential price, not their service.[59] In order to secure ransom, it was obviously imperative for the Africans to keep the Europeans and North Americans alive. And the latter likewise knew that they would die of thirst or hunger without their captors. Of those sailors who wrote testimonies, some spent just a matter of weeks between capture and ransom at Mogador, while others, such as American Robert Adams, who was wrecked on the *Charles* on El Gazie beach in 1810 while sailing to the Isle of May for salt, spent several years as a captive, was passed between captors, and traveled as far south as Timbuktoo before eventually being ransomed at Suerrah (Mogador).[60] Likewise, sixteen-year-old Liverpudlian Alexander Scott was left stranded somewhere between Capes Noon and Boujdour in 1810 after traveling from Brazil on the doomed *Montezuma*. He was held captive for six years and allegedly descended as far south as the river Niger.[61]

On their journeys, sailors suffered greatly from hunger and, above all else, thirst. The hot sun exacerbated the latter and comes to dominate the narratives once the accounts of journeys by sea make way for accounts of captivity in the desert. The sun scorches the shipwrecked, intensifies their extreme thirst, blisters their skin, and generally causes misery throughout their time in the desert. Sun means danger. The shipwrecked conjure its lethalness by connecting it to "the savages'" weapons via a description of the play of light on metal.

Cochelet says from the coast near Cape Boujdour, as he witnesses troops of Bedouin Arabs approaching, "The brilliant display of their arms, which were reflected by the rays of the sun, soon discovered them at a distance."[62] Brisson recalls his first encounter with "the savages" in a remarkably similar way. As the Europeans and Saharawis advance towards each other, Brisson realizes the latter are armed only at the moment when "the rays of the sun [a]re reflected from the polished steel of their daggers."[63]

Inland, the wind becomes a mere daily menace, only on occasions becoming life-threatening. At sea, wind is mariners' tool. Knowing it, and using it to their advantage, is the essence of sailors' craft. In the desert, all the sailors can do is attempt to defend themselves against the cold of the wind. Wind is sparsely referred to, and with little variety in adjectives (*high*, *cold*, *fresh*, or *strong*—rarely *fair* or *fine*, as it is at sea). In Paddock's account, for example, his fifteen mentions of *wind* in his first chapter, set at sea, are accompanied by a variety of adjectives and evocative prose. Wind has a gender and agency. She is alive, vivacious, and volatile—"suddenly" changing in mood and direction. In the remaining fifteen chapters, there are but twenty-three mentions of *wind*, mostly evoked to discuss how Paddock and his comrades attempt to shelter themselves from "its" (the element loses gender and, with it, humanity) cold. "The wind being high, and the night very cold, we suffered much"[64] is a complaint that Paddock repeats with very little variation in wording: "The wind blowing strongly all night, we were extremely cold"[65] and "What with meeting a cold north wind, the fatigue and chill was such as I could no longer bear."[66] The only occasion on which Paddock uses a more detailed and empirical language to write of the wind during the captivity period is upon spying a harbor from the desert. He comments, "From the windward side of the harbor a ship might lie out very well, with the wind as it then was, which blew strong four points on shore, or at north-east."[67] The juxtaposition of his precise knowledge of the wind when it comes to sea-bound navigation compared to his lack of the same on land reveals a fault line in colonial presumed expertise.

Other sailors comment on the wind in similar terms to Paddock while inland. At sea, sailors record their decisions with confident phrasing, showing their ability to master the (often gendered) wind. Captain Riley, for example, tells us, "I came to a determination to haul off to the N. W. by the wind at 10 P.M. as I should then be by the log only thirty miles north of Cape Bajador."[68] In the desert, the wind and sun change the balance of power, and the sailors' phrasing becomes less confident. Wind loses its gender and becomes an *it*, and sailors are at its mercy. After a long day's march, Hector Black states, "Pierced by cold; [I] pulled up some shrubs or bushes to make a shelter from the wind."[69]

Riley also repeatedly complains of having to sleep on the ground while the cold wind "blew a violent gale."[70] Paddock likewise frequently complains of having to sleep in the "cold wind" with "no bed but the sand."[71] Wind's temperature is almost always described at the extremes of the hot/cold binary. At night, it is chilling. By day, wind burns like air in an oven. Sand, a windblown element in the desert, scorches mariners' feet, irritates their skin and eyes, and makes for an uncomfortable bed. Wind and the windblown inform a particular touchscape of the desert, and it is a touchscape of extreme hostility.

However, Riley also remarks at how his captors make use of the wind to hunt ostriches. In great detail, and evidently impressed by the Saharawis' art, he describes how the hunters bid their horses windward, forcing the ostriches to run against the wind and thus lose their ability to use their wings and being careful to ensure they use the direction and force of the wind to leave the ostriches eventually breathless.[72] Saugnier also describes in some detail how the "wandering Arabs" on horseback used wind to hunt ostriches.[73] The Frenchman also comments on the Saharawis' ecologically aware practices such as their insistence on only using dead wood for kindling and never any live shrubs.[74] There are several moments when sailors halt their general tone of disdain for their "barbaric" and "inferior" captors and remark, with admiration, at their skills and knowledge. Paddock, whose narrative becomes dominated by scorching sun and incredible thirst once on land, repeatedly marvels at the Saharawis' ability to locate water. On one occasion, he describes how five of the tribe "went about a hundred yards from us, fell to work under the edge of the bank, dug out an old well, and in less than an hour found water. It had been partly stoned up, but was now completely filled with sand, leaving no appearance, that we could see, of its having ever been a place where water could be found."[75]

Brisson travels to retribution in Mogador with Sidi Salem, the "kind" and "generous" old Arab who almost uniquely escapes Brisson's accusation of all Arabs of being "savage monsters." He wonders at Salem's navigational skills. Brisson seems nervous that Sidi Salem might lose his way as they headed to the coast. "Make thyself easy," says Salem. "The sun is my guide, and will direct me truly." Brisson asks the reader to "judge [his] pleasure and astonishment, when after two days journey [he] found [him]self on the sea-shore."[76]

The epistemological politics are palpable. Faced with this superior Saharawi knowledge, the shipwrecked employ strategies to attempt to evidence their navigational prowess and ability to read the sun in the desert. The example of New York sailor Robert Adams is a fitting one. His testimony is penned by Simon Cock of the African Office of the Committee of the Company of

Merchants Trading to Africa. Cock interviewed Adams in London, where the sailor had returned after having been ransomed by the British Consul in Morocco. Cock assesses in the narrative—usually confirming by way of his allegedly superior knowledge of the region's geography/history—the veracity of Adams's version of events. Cock states that the primary aim of recording Adams's testimony was to provide useful information to the government in its contemporary "exploratory expedition to Africa."[77] There is scant mention of the wind in the two-hundred-page volume. However, the reader is given the impression that Adams may have talked to Cock of the menaces of the desert wind far more than the text indicates. Cock explains that he has omitted discussing "the dangers" incurred from the wind since "they have been so fully described in other places, that any further detail here would be unnecessary."[78] Cock does occasionally allow the inclusion of Adams's heliocentric sufferings, noting how the poor sailor's skin blistered under the "scorching sun" and that he was forced to dig holes at night to escape the heat while trying to sleep.[79] However, most mentions of the sun, which are ample, are included in order to argue for the veracity of Adams's account. For example, explaining why Adams can give, from memory, the length of journeys from various locations around modern-day Mali, Mauritania, Western Sahara, and Morocco, Cock explains, "Being obliged to travel almost naked under a burning a sun, he always inquired, before setting out on a journey, how long it was expected to last. In the progress of it he kept an exact account; and when it was finished, he never failed to notice whether it had occupied a greater or lesser number of days than he had been taught to expect, or whether it had been completed exactly in the stated time."[80]

Likewise, Adams reported to Cock his reckoning of distances and courses between the different places where he stayed during his captivity. Cock was astonished at the accuracy in Adams's estimates, since they were made "with no other compass than the rising and setting of a vertical sun."[81] Cock explains how Adams used the sun to determine direction:

> He always noticed in a morning whether the sun rose in his face, or not: and that his thoughts being forever turned to the consideration of how he should escape, he never omitted to remark, and as much as possible to impress on his recollections, the course he was travelling, and had travelled, and to make inquiries on the subject. Being a sailor, he observed, he had the habit of noticing the course he was steering at sea; and therefore found no difficulty in doing so, when traversing the Deserts of Africa, which looked like the sea in a calm.[82]

Cock thus implies that Adams was able to apply his seafaring knowledge to aid in navigating the desert. To claim ingenuity in navigating by observing where the sun rises and sets and by asking the captors how many days it takes to travel from point of departure to destination seems a desperate grasp at reasserting European scientific superiority. Nautical expressions that travel into the testimonies of captivity play the same role. Mariners, in their testimonies, use abbreviations of the compass rose, as they would apply to wind in their logbooks, when describing winds and their direction of travel. Hector Black, for example, writes that his party headed "in a N.E. direction."[83] Adams and Riley, likewise and in turn, use practical terminology, normally reserved for seafaring, in their descriptions of travels through the desert. See, for example, Riley's observations on the movements of a locust swarm: "The locusts had attempted at night to migrate across the straits into Spain, flying before the wind, which was fair, and blowing from the southward; but that they were either lost in the fog, or checked on their passage by contrary wind, (which generally prevail in the straits at night, particularly in the summer time)."[84] The nautical terminology applied to wind—fair, contrary, prevailing—and his observations on its direction add a tint of *Western expertise* to the narrative. In the desert, far from their nautical toolboxes of chronometers, compasses, maps, protractors, papers, and pencils, the sea captains had nothing. No useful knowledge. No navigational ability. No civilized superiority. Extremely vulnerable, and in denial of their complete dependence on the "barbaric" natives, fleeting recourse to seafaring terms in the narratives of the shipwrecked constitute failed blows in an epistemological battle over the Saharan winds.

Wind and Sun Pathologies, Psychologies, and Physiologies

In 1800, Captain Judah Paddock and his ship, the *Oswego*, found themselves quarantined in Cork due to the smallpox infliction of one of his crew. Blinded and covered head to toe in mange, the poor fellow lay wrapped in a blanket in the steerage, unable to get up. Captain Paddock was reluctant to carry him out to see a doctor for "the danger of exposing him to the cold N.E. damp wind."[85] Thankfully for Paddock, a doctor eventually stepped aboard, the suffering crewman made a full recovery, and Captain Paddock was free to steer the *Oswego* toward the "Cape de Verd" islands. There, he intended to procure salt and skins to take back to New York.

For Paddock, wind had pathological qualities. If he had not been careful, he believed, the northeast wind could have finished off his crewman. Although it has received little scholarly attention, meteorological pathology was part and parcel of the colonial project. Some colonies were thought to offer restorative climates (for example, hydrotherapy or warm weather to cure illnesses), and others, such as the tropics, were considered pathological, especially to Europeans.[86] In the mid-nineteenth century, the colonial powers associated the African coast with sickness.[87] Above all, West Africa, and its winds and airs, were associated with disease and death.[88] Western Sahara's climate fit neatly into the morbidly pathological category.

European colonialists have a history of intertwining climatic and racial ideologies to create ethno-climatological frames of reference for discussing the colonies and their Indigenous people.[89] Referring to the torrid zone, Azurara remarks that the "scorching" and "strange" heat of the sun is the cause of Africans' black skin.[90] Blackmore notes a precursor to Azurara's work in the shape of Alfonso X's translation into Spanish of an Arabic astrological treatise by 'Ubayd Allâh al-Istiji, completed in 1259. The book invokes medieval physiology and psychology to argue that African temperaments, capabilities, and physical features were conditioned by the extreme heat of the sun, which made Africans slow in movement, lacking in intellectual ability, disinclined to pursue science and knowledge, lawless, dark of skin, and "woolly" of hair.[91] Much later, in his account published in 1792, Brisson likewise insisted that Saharawis had a reptilian skin, induced by the climate. He noted the effects of the ferocious sun on his own skin, commenting that he had twice shed it "like the poisonous reptiles of this inhospitable climate" and found that he had developed a layer of "scale, resembling the natural coating of the Arabs."[92] As well as possessing scaly skin, Brisson tells us that Saharawis are "strong and vigorous," have long beards, "large hanging ears," and nails so long that they appear to be "claws."[93] He effectively describes a less-than-human beast evolved to survive in the desert climate, which is convenient for his repeated accusation that the Arabs that kept him were nothing but "savage monsters."

The shipwreck and captivity narratives of Western Sahara bring together the concepts of meteorological pathology, psychology, and physiology. European and American sailors used a reverse pathetic fallacy to make the Indigenous Saharawis as cruel, threatening, and illogical as the Sahara winds and sun and as inhospitable as the "dreary" desert land in which they lived. It is often taken for granted that humans are separate from nature, but this dualism has its origins in Descartes and the enlightenment. "Culture" is often associated with human artifact, an agent that strives for domination over nature, or the embodiment of human rational action that can master and overcome the chaos of nature.[94] To regard the Indigenous people of Western Sahara as part of nature was and is therefore to deny their humanity. As Sandra Ponzanesi shows, European colonials portrayed natives as closer to nature, and therefore at an earlier stage on the evolutionary scale—unpredictable and irrational.[95] Sailors in their testimonies discursively, if possibly subconsciously, linked the "cruel," "wild Arabs" to the equally cruel sun and wind.

A contemporary history of European voyages and discoveries in Africa, published in Brisson's lifetime and drawing on his narrative, suggests that the French diplomat's descriptions of his captors was considered a caricature of exaggerated cruelty even in its time. The nineteenth-century historians explained of Brisson's account the "appearance of insensibility which misery produces, assumed the form of deliberate cruelty."[96] Christine Sears has noted that enslaved Europeans and Americans often remarked on the Africans' cruelty in not giving them much water, but that the former failed to note that the Africans probably had no water to give, or were rationing it.[97] A quotation from Paddock's account illustrates this well: "We lay groaning and crying out for water, and at the same time our limbs were in excruciating pain from fatigue: the merciless barbarians then gave what remained in the skin."[98] Despite knowing that the Saharawis had given the last of the water rations to him and his men, Paddock is only capable of perceiving cruelty in their actions. He, and several others who were shipwrecked, repeatedly remark at the barbarity of their treatment at the hands of the Saharawis in being made to sleep without shelter from the biting-cold night wind or on waterlogged sands after rain, yet no comment is made on whether the Saharawis themselves had shelter. Furthermore, Robbins, Riley, and Adams all slept inside the tents of at least one of their captors at regular intervals. Likewise, Irving, who repeatedly complains in his journals of the tyranny of the "copper colored naked savages" that "enslave" him,[99] appears to be fed and watered whenever he asks and is even permitted his "enslaver's" place on a camel when he becomes exhausted from walking.[100]

Complaints of cruelty at "refusing" water, food, and shelter are accompanied in the narratives with repeated comments as to the Saharawis' natural or learned resilience to the hazardous wind and sun. This juxtaposition serves to suggest that the Saharawis are somehow conspiring with the local weather to worsen the suffering of the Europeans and Americans. Adams suggests that Saharawis bear the sun better than Europeans,[101] while Riley explains that Saharawis were used to not drinking much water.[102] The latter may have been true, but the implicit suggestion was that the Saharawis were cruel not to furnish their captives with a greater share in the rations (whether they had more rations to share is not considered). Paddock says, "The Arabs, from habit, could go a long time without water, and did not then appear to suffer at all in comparison with what we endured."[103] However, Saugnier's perspective is exceptional. He comments, "Whatever losses an Arab may meet with, he is never heard to complain; he rises superior to poverty, supports hunger, thirst and fatigue, with patience, and his courage is proof against every event. God will have it so, says he."[104] Saugnier suggests that it was Saharawis' courage, their psychological strength, and their outlook that made them more resilient, not innate, "natural," physical characteristics, which again paint the Indigenous as "closer to nature."

Psychology comes into play not just in bearing difficult weather but also in how the weather affects one's emotional self. Our emotions also prompt how we perceive the weather and our wider surroundings. Context, of course, plays a vital part. A storm experienced at sea is not the same as a storm rattling at the shutters of a cozy, fire-lit living room. Climate in the Sahara Desert threatened the shipwrecked in material, physical ways. But the emotional state of the shipwrecked further tainted their perceptions of the desert weather-world. We can observe this by the shipwrecked descriptions of the desert, which is a hellish place at their lowest emotional moments, but which offers certain charms when the sailors were of higher spirits.

Paddock's first impression of Western Sahara, upon ascending a dune from the shore at dusk, was of "barren sand, without either tree or shrub, or the least appearance of vegetation; dreary in every respect."[105] Cochelet repeatedly describes the territory as "dry and barren,"[106] and Adams describes it as "barren and uninhabited."[107] Hector Black witnesses only "sands and barrenness" with the occasional shrub or bush tempted up by dew in order to allow him "to make shelter from the wind."[108] But then as salvation draws near, the landscape becomes more varied and beautiful. Brisson, for example, finds himself at the start of his captivity in a land of burning-hot pebbles as white as snow, as flat as lentils, with "no variety," "not producing any plant whatsoever," and with the horizon obscured by red vapor "as if there were burning volcanoes on every side."[109] He describes a dreadful silence, with no bird or insect to be seen, and remarks that should a breeze arise, the traveler's "lips crack; his skin is parched up, and little pimples, that occasion a very painful smarting, cover his body."[110] The wind is responsible for a pathological touchscape. The perception of Western Sahara's soundscape as a "dreadful silence" has parallels with the colonization of Australia, where early colonists interpreted the local soundscape as a "great silence."[111] Accounts of the desert's absolute silence in Australia and Western Sahara contributed to the construction of an aural terra nullius. Nevertheless, when Brisson's "cruel" owner passes his custody to kindly Sidi Salem, who Brisson is sure will take him to the consul in Mogador, his eyes open to "an astonishing variety of scenery."[112] Pages of his narrative are dedicated to his wonder at "mountains of prodigious height," valleys of date palms and laurel, forests, peaks with pebbles of "red, yellow, blue and green," and, as they approach the lands of the Tekna tribe nearing Morocco, hamlets with relatively abundant water sources and barley fields.[113] Riley too, who for the most part describes the territory as "barren," "dreary," and "inhospitable," sees the territory and indeed experiences its nightly wind, which had previously so harassed him, in a different way upon achieving hope of redemption. After forming a friendship with his captor Sid Hamet and trusting the latter will take him to Mogador for ransom, Riley delights in the desert one morning after a storm:

> It had rained very hard with heavy squalls of wind most part of the preceding night, but my tent being sound, kept off the storm: it was now clear and serene; nearly the whole face of the ground was covered with violet and pink coloured flowers, not more than an inch or two in height, which seemed to have sprung up during the night, and as the sun exhaled the dews from around them, the fresh air of the morning was filled with the most delightful fragrance.[114]

In this example from Riley's narrative, we see that the violent night wind outside gives meaning to the tent, while the sea captain's better emotional state opens his eyes to the small beauties to see and smell in his environment. Misery had previously blinded him. As Alexandra Harris has pointed out, when a joy absorbs us, we are more likely to see the weather in a positive light. In times of vulnerability, the opposite is true.[115]

Stormy winds, though, also offer a different lesson in meteorological psychology. One wind in particular strikes terror in the shipwrecked. It is a hot wind that blows in from the east, carrying enormous quantities of sand. Their contemporary, Jackson, the European merchant living in Morocco, calls this wind the *Shume*. Spanish colonials later and Saharawis today call it the *īrīfī*. According to Jackson, the Shume is a "tremendous," "hot," and "impestuous wind" that "seldom or ever continues more than three days."[116] He goes on with the following: "The heat is so extreme during the prevalence of the Shume, that it is not possible to walk out; the ground burns the feet; . . . the parching heat of the wind . . . resembles the heat from the mouth of an oven."[117]

With the Shume/*īrīfī*, wind and solar heat combine to form the climax of nature's chaos in its Saharan form. In terms of the aftereffects of the Shume wind, Jackson finds it converts the desert "into a moveable sea, aptly denominated by the Arabs 'El Bahar billa maa,' a sea without water, more dangerous than the perfidious waves of the ocean."[118] Such a sea of sand is experienced by Charles Cochelet, whose captors, standing on a precipice before a long stretch of sinking sand that they must cross—a Sisyphean task—begin to pray for guidance, which leaves quite an impression on the sailor: "Their accustomed prayers acquired, in this conjuncture, a new degree of fervor, and to these were added various spiritual songs, which they sang in a loud and piercing voice, and with extraordinary volubility. The image of these two men, prostrate and trembling upon the sands, imploring a passage across this chaos, which represented to me that of all nature, will never be effaced from my memory."[119]

If wild, untamed, hostile nature was something for the Western traders and sailors to seek to master and tame, the Shume/*īrīfi* was—as Cochelet hints—the ultimate challenge. Riley describes his experience of such a wind near Cape Bojador: "Immense heaps of loose sand, forming mountains of from one to three or four hundred feet in height, blown and whirled about by every wind, and dreadful to the traveller, should a strong gale arise whilst in the midst of them; for he and his beasts must then inevitably perish, overwhelmed by flying surges of suffocating sand."[120] Paddock experiences the Shume and expresses his wish to die: "The sun was so powerfully hot, that the sweat not only dropped, but on most our faces it ran a stream. We were permitted to lie down a while, and soon fell asleep. The wind blowing fresh, we were very soon called up, when we were more than half covered with drifting sand, and no doubt in a short time we should have been buried alive in the drifts. I believe there was none among us, but would have been willing to remain undisturbed, and die there."[121]

Archibald Robbins, upon his own encounter with this "violent" wind, likewise expresses his closeness to (desire for?) death after losing his senses to the blowing, hot sandstorm: "My ears, and nose, and sometimes my mouth, were literally filled with sand—that one almost lost the sense of hearing—the other that of smelling, and the last that of tasting. . . . A few more such day's travelling, I think, would have put an end to my life and my slavery."[122] Although the *īrīfi* was the utmost climatic indicator that Western Sahara was savage and needed to be civilized, it also appeared as totally indomitable and constituted the sailors' moral defeat, if not quite their physical one. If wind is central to the sailors' perceptions of Western Sahara's touch and soundscapes, and these 'scapes, in turn, were vital to the construction of Western Sahara as terra nullius, Robbins's alleged loss of hearing, smell, and taste at the hands of the wind and windblown is surely the superlative of a sensory terra nullius and hostile, wild, and barbaric sensory place.

Conclusion

Wind and sun were the principal obstacles to survival for those shipwrecked on or near Cape Boujdour. The Indigenous people were not a threat to life for those taken captive. On the contrary, the sailors' survival was crucial for securing ransom money in Mogador (Swearah). Nonhuman animals were likewise not mentioned as a fear. But the intense solar heat—accentuated by the wind in daytime—the lethal sandstorms brought by the *īrīfī*, and the thirst and hunger that accompanied the hot climate constituted challenges of mountainous proportions. The desert weather-world setting for the narratives is outlandishly inhospitable and brutal, but this desert weather-world is more than just a setting; it is a character and a subject.

As well as tormenting them physically and emotionally, the extreme heat and extreme wind also served useful purposes for the shipwrecked. Wind and sun are used as metaphors for the barbarism of the Saharawis, who are generally painted—with generous recourse to superlatives—as inhumanly ferocious. The hyperbolic descriptions of the hostility of the weather and desert sands is used to accentuate the alleged cruelty and savageness of the Saharawis. The shipwrecked also painted the Saharawis as somewhat weatherproof: comments on the ability of the Indigenous to better bear thirst, intense solar heat and wind, and lack of food and shelter further dehumanize the Indigenous population.

The basis of the colonial enterprise was that Europeans had a divine duty to civilize wild people and wild nature. Sailors had mastered wind and used it to trade, enslave, and colonize around the world. Therefore, to be shipwrecked and completely reliant on their captors for survival was a mark of the limits of European or American colonial mastery of the wind. Sailors' use of logbook language to describe the winds in their narratives is an attempt to mask their lack of knowledge of Saharan winds, desert navigation, and survival.

Although some shipwrecked mariners highlight the kindness of individual Saharawis, and even claim friendship with them, in general, the Saharawis, as a people, are depicted as clawed monsters evolved to dwell in the horrible, dreary, barren desert. Like the environment where they live, the Saharawis are inhospitable. Like the menacing desert wind and sun, they are cruel. As is typical in colonial discourse, painting the Indigenous as barbarians justified a civilizing mission, as did painting natives as closer to nature. Likewise, although in fleeting moments when their spirits were high, the shipwrecked appreciated the beauty of flower blossoms after storms, date palms, and awe-inspiring mountains, their general description of the desert amounted to a barren, hellish, empty place. Again, this is well within the imperial tradition.

Imperial powers treated colonized lands as vacant spaces to justify natural resource exploitation and total disregard for Indigenous constituents. Although the shipwrecked found themselves in Western Sahara by unfortunate accident rather than with the intention of colonizing the land, their narrative depiction of the desert weather-world and its human inhabitants laid the groundwork for future colonizers. Their Saharan wind imaginaries, along with all their political implications, would inform how colonizers-to-come would understand Saharan winds, lands, and people, and, in turn, shaped the Spanish colonizers' energy developments.

CHAPTER 2

The Red Wind

Aeolian Anxieties and Energy Infrastructure in Spanish Sahara

---

I suppose this chapter should start with a white fortress, *la factoria*, on Western Sahara's coast, not with the Red Wind, as the title might suggest. *La factoria*, the prize of Emilio Bonelli y Hernando, first governor of Spanish Sahara, with his meandering W-shaped moustache, military badges, tea set, and kif pipe.[1] *La factoria*, a typical Spanish army fort, a forty-to-sixty-m^2 rectangle, with stone battlements and masonry embrasures through which to shoot, its canons and machine guns poised ready.[2] *La factoria*, constructed as the Age of Sail of Captains Riley, Paddock, or Irving ended and the Age of Steam, with its faster vessels that no longer relied on the winds, ascended. *La factoria*, a little south of Cape Boujdour and all its ship and sailor skeletons, all its captive Europeans and North Americans. *La factoria*, a fort that served, rather ambitiously, as the "capital" of the colony,[3] the only Spanish building in the country for the first years of Spanish Sahara's existence. *La factoria*, the seed from which Villa Cisneros city grew, the last urban center to be abandoned by the Spanish upon the Moroccan invasion.[4] *La factoria*, which ceased to exist in 2005 when the Moroccan state bulldozed it. *La factoria*, once powered by renewable energy. *La factoria*, haunted by the wind.

During the European Scramble for Africa, Spain was attracted to the lands surrounding Cape Boujdour and beyond partly because of Canary Islander fishermen. The latter had fished and hunted whales and seals off the Sahara's coast since the 1500s. Spain wished to provide protection and safe ports for the Spanish islanders, as well as for commercial steam vessels traveling to

Equatorial Guinea, especially given the cape's history of wrecks. Spain's *la factoria* was an initial modest claim on the Saharan territory.

In the early days of its existence, a whooshing and screeching wind stormed the fortified factory night after night, plunging its resident soldiers into darkness, shattering all the lamps, and impregnating the air with the latter's petrol scent. To end this problem, in 1913, electrical engineer Jose Galvan Balaguer was tasked with electrifying the fort. He wrote, "The wind that constantly reign[ed]" in Villa Cisneros, with an "excessive" speed of twenty four kilometers per hour for most of the year, "ma[de] the lighting of the building by petrol, the only source [then] in use, almost impossible."[5] Galvan Balaguer designed an electricity-generating wind turbine for *la factoria*, which would avoid the dangers of petrol lamps and would be cheaper and easier to manage than a gasoline-powered motor and dynamo.[6] Thus, the first electrical system designed for Spanish Sahara was to be run on moving air, a practical solution to challenges posed by an "excessive" wind.

How one comes to know the wind of a place is never politically neutral. In his writings, Galvan Balaguer personified the Saharan winds and regarded them as a sovereign power. It was as if wind resisted the colonizers. Like a poltergeist, wind came at night and terrorized the men of *la factoria*, throwing things and smashing them against the walls. Below, I focus on the writings of the Spanish men that followed Galvan Balaguer and the soldiers at the fortress to Spanish Sahara, arguing that the Saharan winds haunted colonizers throughout the colonial period. Wind exposed the limits of the colonizers and their technologies, thus provoking deep-seated anxieties. These aeolian anxieties shaped Spanish colonialism. On the one hand, Spanish writers imagined wind and windblown desertscapes in a way that attempted to justify the colonial project and to colonize Western Sahara and its people at the imaginary level. On the other hand, colonialism responded to the winds' threats and resistance. In both ways though, and as I argue in the second half of this chapter, colonial wind imaginaries shaped the development and management of energy infrastructure in Spanish Sahara.

The Scientific Colonization of the Sahara

Spain's first steps into the European Scramble for Africa preceded the Berlin Conference by mere months. They were initiated by the Geographical Society of Madrid. The society, created in 1876 by military engineer and geographer Francisco Coello, promoted exploration, travel, and colonial projects in Africa. A year later, Coello founded the Spanish Association for the Exploration of

Africa, which aimed to explore the rich fisheries off the coast of Western Africa.[7] In 1884, just before the Berlin conference and partly in preparation for it, the Spanish government, in partnership with two Spanish companies, asked the Arabic-speaking Emilio Bonelli to head an expedition to the Sahara.

In 1884, Bonelli thrust the Spanish flag into Saharan sand. As first Governor of Villa Cisneros, he built the aforementioned fortified factory from which to trade and planned expeditions to survey the territory with a view to determining its economic possibilities. He took part in an initial expedition near the coast but then commissioned two locals to carry out expeditions into the desert's interior. The second major Spanish expedition was sponsored by the Spanish Society of Commercial Geography in 1886, led by Commander Julio Cervera, Commander Álvarez Pérez, Geologist Francisco Quiroga y Rodriguez, and Arabist Felipe Rizzo. Full of daring-do, their expedition took them 425 kilometers inland along the Tropic of Cancer to a place known as the *Sebja* of Iyil, modern day Mauritania, where they signed commercial agreements with the local tribes. The Spanish hoped to intercept the caravan route that passed between Timbuktu and Morocco. The Sebja, where caravanners exchanged their wares for salt, seemed the ideal place to do so. Maintaining Spain's access to rich fisheries was also an attraction.

Ultimately, Bonelli's economic and political ambitions were thwarted by repeated attacks from Saharawis. On one such occasion, six Spanish men died and several more were injured and subsequently evacuated to nearby Las Palmas in the Canaries. Humiliated, Bonelli returned to the peninsula and, in 1904, his replacement General Francisco Bens arrived. Bens was astonished to learn that not only did the Spanish employees of the Bonelli Fort have to pay tribute to local Saharawis, but also that the only territory of Spanish Sahara that the Spanish actually controlled was the land on which the fort was built. But Bens had a strategy. A skilled diplomat, he remained in the Sahara as governor until 1925 and was largely responsible for Spain's gradual extension of influence in the desert. He managed this, explains San Martin, by building solid bridges of trust with the Saharawi *qabā'el*.[8] He learned *Ḥassānīya*, mixed with everybody, practiced Saharawi laws of respect, and extended hospitality to Saharawi leaders. Cuban-born Bens was seen as "a living myth in the desert" and allied himself with the *qabā'el* against French and Moroccan threats in exchange for safe access to Western Sahara's coastal fisheries for Canary Islanders.[9]

It was not until the 1930s and 1940s, after Bens had slowly and tentatively established a few garrisons across Spanish Sahara, that the Spaniards were able to explore "their" new territory.[10] Shortly after the end of the civil war, the

Franco administration created the African Studies Institute (IdEA). Between 1941 and 1947, IdEA sponsored several expeditions with the aim of generating knowledge of Spain's African colonies. One of the priorities of 1940s Francoist Spain was to seek economic profit from its African territories, and the possibilities of Spanish Sahara, apart from the fisheries, were still largely unknown at that time.[11] The flurry of Spanish geographical, environmental, anthropological, zoological, geological, and sociological studies in this period, some of which are referred to below, thus had economic ends. As Benita Sampedro Vizcaya points out, such studies can be understood as a form of appropriating the natural world ahead of its occupation.[12] Among other outcomes, these 1940s expeditions and research projects resulted in the discovery of phosphate and oil deposits in Spanish Sahara. However, the economic and political challenges of the post-war period in Spain and the Second World War in Europe stalled further exploration until 1960.[13]

Exploration in the 1960s brought networked energy infrastructure. General José Díaz de Villegas, the director general of Spanish possessions in Africa, explained his rationale for the construction of electrical production and distribution installations in El Aaiún, in 1960, as follows: "The new activities that are being carried out in the Sahara Province have triggered an extraordinary influx of people to the capital, related to the oil studies or to the linked and complementary industries to serve these [oil exploration activities]."[14] Electrical communications were likewise introduced in the interest of oil exploration. Communications engineers in Madrid drew up the first plans for a telephone service in the Sahara in 1961. It would be connected, at both El Aaiún and Villa Cisneros, to Las Palmas, Gran Canaria. The engineers commented, "Of course, it's obvious that the telephone traffic as well as the telegraph will depend on the development and result that is obtained with the oil explorations that are ongoing in the Sahara. However, it's doubtless that before waiting for those results, it's necessary to get electrical communications, precisely in order to contribute to the development of said exploration."[15]

While Spain developed networked electrical infrastructure to service the oil industry, it was the discovery of phosphates that brought crucial changes in the shape and intensity of the Spanish colonial project. Alicia Campos Serrano and Violeta Trasosmontes conceptualize this development as leading to a "second occupation" of Spanish Sahara,[16] while Lino Camprubí finds that the resulting development of the mines and phosphates exports constituted a "technological transformation of the Saharan human and physical landscape from a desert land populated by nomadic groups into a militarized phosphate production facility."[17] Almost a third of the entire Spanish army, which constituted three times the adult Saharawi population, was stationed in Spanish Sahara, creating, in Pauline Lalutte's words, "a vast military garrison" with the aim of "protect[ing] the phosphates."[18] Oil exploration activities in Spanish Sahara coincided with phosphates prospecting. While oil companies did not employ any Indigenous locals, the phosphate mines relied on Saharawi and Moroccan labor (Saharawis made complaints that jobs in the phosphates industry went to Moroccan immigrants, or to immigrants posing as Saharawis).[19] After oil companies left in 1963, due to more lucrative opportunities in Libya and the North Sea and a global fall in petroleum prices, phosphates became the central focus of Spanish energies. Spanish Sahara's phosphates were high quality and relatively easy to extract, and the Spanish-built mine at Bucraa, near El Aaiún, became the world's largest open-cast phosphate mine.[20] If oil exploration had caused the rapid growth of the Spanish population and infrastructural developments in the colony, the discovery of phosphates ballooned it and, moreover, saw the incorporation of thousands of Saharawis into the sedentary urban population and workforce.

Colonial research expeditions and networked electrical infrastructure both served natural resource exploitation. The publications resulting from said expeditions, as well as from pseudoscientific writings and memoirs from military personnel and other colonial actors, shared something else: a major preoccupation with the wind. Wind—"the most prominent factor of the desert" in botanist Emilio Guinea López's words—and windblown desertscapes have a strong presence in all colonial scientific and pseudoscientific writings on Spanish Sahara.[21] These scientific studies were limited, and therefore also shaped, by the Saharan winds themselves—most importantly the *īrīfī*—which halted the researchers' movements, delayed their schedules, and would have endangered their lives if not for Saharawi guides. Saharan winds were both a source of colonial anxiety, constantly reminding the colonists of the limits of their powers, and a tool to justify and celebrate Spanish colonialism at the imaginary level.

Aeolian Anxieties

The sand blew in such volumes and with such vigor, working its way into every nook and cranny with such precision, that it broke Señor Quiroga's compass and lodged itself between the glass and the face of his aneroid barometer. The menacing wind and its aerosols (sand and dust swept up by the wind) destroyed his ability to scientifically know the desert climate and were a constant barrier to progress during his 1886 expedition. For Europeans, meteorology, geology, and geography were all methods for coming to know the colonies and to facilitate life for settlers in "their" new lands.[22] As Malony and Enfield point out, early colonialists from across Europe—missionaries, soldiers, medics, explorers—packed barometers, thermometers, and other bits of meteorological kit in their trunks; measuring the weather was a key practice of colonialism. It made foreign environments legible to European sciences and states.[23] Sampedro Vizcaya likewise asserts that mapping a colony's weather conditions was a "critical first step to exploration and settlement."[24] Given their alleged intellectual inferiority, natives were not believed to be able to know, map, and understand nature in the same way that "civilized" occidental society allegedly could. Quiroga's meteorological tool kit was meant as one more way to prove Western superiority. The shock was that his tools, in the desert, were not fit for purpose.

Quiroga's early experiences were relived by later military officers and scientists. In his autobiography, Ignacio Hidalgo de Cisneros, for example, writes of his time as commander of the Spanish Sahara's air forces from 1927 in Cabo Juby and Villa Cisneros. Cisneros, who would later become commander of the Republican air forces during the Spanish Civil War, lived with other aviators including Antoine de Saint-Exupéry on the coast of the Sahara. For him, the Spanish Saharan touchscape was characterized by menacing windblown sand. The winds and the sands were a "nightmare."[25] He describes how sand floats permanently in the air, getting into his food, his clothes, and his hair. Harassed to delirium, Cisneros shaves his head. But the wind still mocks him and ogles at him from outside the window. Overnight, says Cisneros, a sea view would become blocked by a giant harrowing sand dune, and it was a constant fight not to be buried alive. He tells us how he could never have imagined that the wind would blow "day and night, weeks and months without stopping for an instant."[26] His "greatest worry," though, was not for his personal well-being but for the aerodrome's apparatus.[27] The carburetors kept breaking with the sand, and the humidity corroded the metal parts of the planes. The wind made life a struggle and rendered much colonial technology precarious.

The winds and the aeolian desertscapes did not just expose the limits of colonial technologies at the blustery coast. They also bullied inland expedition members. Eduardo (geologist) and Francisco (geographer) Hernández Pacheco carried out a government-sponsored expedition to Spanish Sahara in 1941 to explore the geology, physical geography, fauna, and vegetation of the areas recently occupied by the army. They soon found that they could not drive anywhere. Attempting to maneuver a car in areas of "miniscule dunes . . . arranged in the direction of the wind" was "annoying and slow" while the larger dunes made it all but impossible.[28]

These are but two examples of a plethora of anecdotes concerning windblown sand-provoked technological and mechanical failures in Spanish colonial nonfiction writings. One (perhaps unconscious on the part of some of the writers) effect of such tales was to aggrandize later infrastructural developments. If the climatic conditions were so hostile as to render useless a compass, car, or carburetor, then successfully developing built or networked infrastructure was a true triumph of colonial modernity and technological acumen. Beyond noting technological failures, though, Spanish colonial nonfiction writers furthered the wider terra nullius narrative developed by eighteenth- to nineteenth-century European and American mariners. They did so through recourse to descriptions of hostile climatic conditions. Terra nullius has its origins in Roman law, in which all lands beyond the imperial boundaries were considered to be ownerless and therefore annexable.[29] In the fifteenth and sixteenth "ages of discovery," colonizers regarded all non-Christian territories as nullius.[30] In the nineteenth-century era of European colonization, the doctrine was used to appropriate the lands of "non-civilized" peoples, who were regarded as incapable of enacting territorial sovereignty and of making effective use (the relationship with capitalism is clear here) of the lands they inhabited.[31] To date though, there is little research into how terra nullius narratives have been applied to wind. Alfredo Eugenio Narbón, lighthouse technician in Villa Cisneros from 1949 until 1976, gives an illustrative example of how Spanish colonizers constructed a windblown terra nullius. In the prologue to his memoir, he offers a dedication to two Spanish engineers that pioneered Spanish Sahara's lighthouses:

> Those trips by camel under the relentless desert sun; those nights out in the open, under the weak tent with the brutal fall in temperature, which made freezing nights of scorching days; those times navigating with a map and compass looking for the paths of future maritime signals; and that risk of the ominous "īrīfī" (the blazing wind of the desert) that can be unleashed at any moment; and the political riots, that even cost the life of some members of our team. . . .
>
> They knew how to face the dangers and difficulties and to press ahead with a brilliant project, which resulted in the Maritime Signals system, which Spain left operational when it left the territory.[32]

Lyall Watson, exploring perceptions of winds in Victorian poetry, found hearty, character-building winds that answered the call of Empire.[33] Such literary winds augmented the chauvinism of British imperialists. Dedications such as Narbón's have a similar effect, inflating the gallantry of Spanish colonists in the face of winds, which are painted as far more bellicose and hostile than the robust winds of Victorian verse.

As was the case in the writings of sailors and other shipwreck survivors, Spanish colonial writers used wind imaginaries to construct Western Sahara as terra nullius. Narbón is appalled by the windblown desert. He finds Rio de Oro, one of two regions in Spanish Sahara (the other was Saguia el Hamra) is "completely flat" with the effect of surrounding Spanish residents with an "inalterable monotony."[34] There is a "blinding" sun that burns without producing changes in tone or color: "It's the same at noon as it is at five, the day is all the same, as are the months and years, succeeding in time without nuance or seasons."[35] In Narbón's descriptions of the desert, wearing adjectives follow bland nouns and morbid verbs. Hyperbole prevails. Such repetition of negatives suggests the desert is most notable for its absences and vacancies. In the Sahara, "man and machine bear wind and sand, heat and dust, lost in the blinding day of the desert."[36] Said wind and sand "run out of control over the dead landscape."[37] To describe wind and sand as "out of control" suggests the need for a (colonial) agent to tame them.

Eduardo and Francisco Hernández Pacheco likewise personified the winds and the windblown with a vocabulary of the pathologically criminal. Of the migrating (untethered by tree or grass) barkhan dune, they said, "It is insensitive and ruthless. In its continuous progression it produces destruction and death. The flowery shrub or the bush covered in buds and leaves that gets in its way, is reached, covered slowly and encompassed in the sandy mass; when the dune has passed, a dead scrubland and dried skeleton of a tree is left behind."[38] For the Hernández Pachecos, the aeolian dune had agency and operated with malevolent intention, smothering all life in its path.

Spanish writers extended their terra nullius narrative to encompass the Saharawis themselves. Doing so created a particular brand of Indigenous savageness. Saharawis became windblown creatures, with a temperament shaped by the climate. This is a manifestation of a wider culture/nature dualism that has been central to colonial encounters (and indeed to Western thought) for centuries. The culture/nature binary is closely linked to others such as civilization/barbarism and modernity/backwardness. The colonizers have almost invariably thought of Indigenous peoples as part of, or at least closer to, nature and *therefore* as uncultured and uncivilized. For colonialists, such imagined Indigenous wildness has justified violently transforming people and ecosystems. Particularities of dualist thought in Spanish Saharan colonial discourse linked the Indigenous Saharawis specifically to aeolian processes. Quiroga's writing provides a strong example. Issues twenty-five to thirty of the *Revista de Geografia Comercial* were dedicated to his aforementioned 1886 expedition. The first page names the factors that made the expedition so difficult and thus Spanish "conquering" of the desert so full of merit: "the harshness of the climate and the barbarism and fanaticism of the natives."[39] The discursive linking of climate and human barbarity is developed throughout the lengthy journal publication, most notably in the section entitled "The Sahara and Its Inhabitants," which embeds a discussion of Saharawis' political structures in a wider exploration of the wind-ravaged landscapes and atmosphere. After spending several pages discussing the wearing effect of the winds on the landscape, and on his own physical well-being (the hot wind completely dries out his corneas), Quiroga comments on the political structure of Saharawi society. His thoughts are worth quoting at length:

> They live in an almost savage state of independence, opposed to all moral and material development. They respect the name of the sultan of Morocco for being a descendent of the Prophet; but they energetically reject the idea of becoming Muley-Hassan's subjects. Their chiefs have no command over the tribe without the approval of all the men that form part of it. The chiefs trust in the council of all members of the tribe, the grandest and richest chief and the poorest camel shepherd enjoy equal freedom in expressing their opinions. Moors from various different tribes have told us a thousand times, and even in front of their own chiefs, that the Arab of the desert has no ruler but "Allah and Mohammed." The Üled-Bu-Sbá tribe, one of the most important in western Sahara due to its size and the commercial spirit of its members, does not currently have a chief.[40]

For the geologist Quiroga, then, the Saharawis' relatively hierarchy-free society of freedom and equality was a sign of their feral and backward nature. His publication is ostensibly concerned with (aeolian) geology. In this context, to juxtapose detailed descriptions of the destructive climate with ones of the wild and backward people is arguably to conflate the two. *Saharawi* becomes a metonym for the terra nullius that is the windblown Sahara—an aeolian terra nullius. Other later writers compare the Saharawis with the aeolian processes of the desert dunes in similar ways. Narbón writes in his memoirs of the Saharawis and their dwellings. The *ḫaīma*, he says, "has the property of changing its position with the same silence and mystery as the dunes."[41] Hence, the Saharawis, their architecture, and their aeolian surroundings share characteristics.

The metonym of the Saharawi for the wider ensemble of a wind-beaten, wild, and savage desert is also salient in Josep Congost Tor's ornithological study of Spanish Sahara. Congost Tor joined a Zoological Museum of Barcelona expedition in spring 1973 to Spanish Sahara during which he studied bird species in and around Smara. Smara was, according to Congost Tor, "occupied by extensive regs almost completely devoid of vegetation and whose only orographic elevations are wadis, monadnocks and krebs."[42] He also asserts that Smara's reg-dominated geology is fruit of the *īrīfi* and describes a dried-up landscape ravaged by wind and sun.[43] His descriptions of the areas in and around Smara are almost uniform in painting a picture of lack: lack of vegetation, lack of water, lack of human habitation, and even lack of significance. This is the familiar European art of defining the desert by emptiness. The description of "urban Smara" is unique in focusing on the human population. It is worth quoting this description in its entirety: "Low-rise adobe houses align wide streets and unpaved squares. Some squares harbour a great quantity of waste and feces. During the daytime, the inhabitants do not frequent the streets and squares. Surrounding the city are extensive desert regs."[44]

There is no context given for this observation of waste and feces in the public square. The feces could be fertile camel dung. Waste could be food for goats. For sure, the Spanish authorities ignored calls from Saharawis to connect Indigenous suburbs to the sewerage system. Any number of explanations is possible, but the reader is handed an image of an utter lack of basic hygiene. To comment on the presence of feces in public squares in the one and only description of how Saharawis lived in urban Smara is to foster a narrative of Saharawi barbarism. Congost Tor's descriptions of Smara and its rural surroundings contribute to the terra nullius narrative and blend the allegedly unhygienic (and therefore implicitly uncivilized) Saharawis squarely into the bland, barren, wind-destroyed wasteland.

Other writers give a less overtly negative view of the Saharawis but nevertheless use them as a metonym for the desert. For example, Eduardo and Francisco Hernandez Pacheco affectionately give the desert as the cause of Saharawis' relatively slow pace of life: they imply that the Saharawi collective temperament is determined by the aeolian space of the Sahara.

> The nomads, with their livestock, in their continuous wayfaring lives, set out on their long travels over the immense plains, towards distant landscapes, where the rain has soaked the countryside and grasses and shrubs are sprouting. They always walk, constant and without haste, because time and desert merge into one same concept of immensity and eternity.

> Without haste and without the worry of arriving late, because the desert, which belongs to no one and everyone, waits, and they will get there in time for the livestock to get fat with the green grasses of the nourishing seeds, for the fruit to mature and the herbaceous plants to dry. Walking slowly in search of pasture, every sunset, in the comfortable *jaima*, a tapestry is laid out, bags of polychrome leather are accommodated, water is boiled for the tea, and everyone rests there until the dawn of another day. Then, the *jaima* will be brought down, the camels will be loaded up, and they will resume their march in whichever direction is convenient. There in the remote distance, when the day of travel is done, they will take their time in setting up the *jaima* once more.[45]

The Hernandez Pacheco duo's observations reflect those of Narbón's on the unchanging monotony of the desert, although they show this monotony in an affectionate, favorable, and poetic light. Yet, the picture they paint shows Saharawis as a part and product of the desert, and their traditions and home as somewhat unchanging: all is still, the sand has no plans. Neither the Saharawis, nor the camels, nor the vegetation have anywhere they need to be. Saharawis—sweet, slow Saharawis. Neither they nor their desert will progress, or modernize.

Biometeorological perspectives, such as the Hernandez Pacheco duo's, also offered building blocks in the construction of a wider windblown terra nullius narrative. Such perspectives created a pathology of place. Geographers have recently begun exploring the early twentieth-century reinvigoration of Hippocratic modes of thinking, which reconceptualized the human body as a biometeorological entity. This popular field of research, shows Livingstone, divided climates into invigorating and debilitating and "thus of feeble and fit people."[46] For Hippocratian geographers, the damning climates of darkest Africa produced the laziest of peoples, incapable of the great accomplishments of the West.[47] Spanish colonizers' writing escapes this most overt racism, which was, as shown earlier, explicit in some testimonies of the shipwrecked. But a similar Hippocratian environmental determinism remains in much colonial Spanish scholarship. While some colonial researchers remarked that Spanish Sahara's heat and dryness killed germs[48] and was favorable to Europeans' health—botanist Emilio Guinea López, for example, commented that "the dry atmosphere submerges the body into a state of physical wellbeing"[49]—the Saharan winds were generally viewed as a pathological threat. In this respect, Spanish writings on the Sahara's winds match that of other European scientists and geographers in finding that Africa's climate made it hazardous for European settlement.[50]

Like the shipwrecked, some Spaniards indicated that Saharawis were naturally able to cope better with the trials of the wind. Observing an elderly man, still straight-backed and agile enough to mount a camel unaided, General Bens remarked on the ability of Saharawis to physically "endure" the wind for decades.[51] Ángel Flores Morales, a topographical deputy on the Geographical Service, offered a similar flattering description: "[Saharawis] are gifted with a marvelous resistance to tiredness and disease. They are strong and elegant up to an old age. Their skin is weather-beaten by the intense winds of the desert, bronzed by the mixture of races . . . their faces emit intelligence and vigour."[52] When reading such descriptions, one gets the impression that Saharawis physically thrive in the desert, as if they were *naturally* suited to the desert ecosystem: a windblown, sandblasted people.

Other Spanish researchers focused on the biometeorological dangers for Spaniards. Manuel Mulero Clemente, for example, was a military commander who published a pseudoscientific book about the territory of Spanish Sahara and its people based, he said, on a review of various published works on the territory as well as on his own notes. Mulero Clemente warned that Spaniards prone to anxiety should avoid the climate of Spanish Sahara, while those with lung disease or rheumatism should specifically avoid the coast.[53] He highlighted several meteorological pathologies of the territory, including those caused by its named winds. The *īrīfī*, for example, is "frequently charged with electricity, and it is normal that men become very restless in its presence, even becoming sleepless the night before its arrival."[54]

According to Mulero Clemente, the *īrīfī* also presented purely physiological, lethal dangers: "The '*īrīfī*,' when it is persistent, represents a grave danger to man in the desert because he is unable to shelter himself or fight the dryness of the wind, which drives the sweat glands of his organism to make an extraordinary effort that, carried to the limit, can lead to death."[55]

Cristóbal Benítez traveled through Spanish Sahara 1899. He was a translator accompanying geologist and mineralist Oscar Lenz on an expedition to Timbuktoo. In his account of his travels in Africa, Benítez says the Saharan winds played psychological tricks on him. As the expedition members pass by some mountains, Benítez mistakes the wind bouncing between the summits for a drum and, on another occasion, the sash-like bands of windblown sand for rising smoke.[56] As the sashes of sand flickered between the summits of four mountains, Benítez wondered if he was seeing volcanoes. On each occasion, an Indigenous guide explained the aeolian and aural processes that resulted in the phenomena that Benítez witnessed. The aural trick suggests the mystery of the windy soundscape for colonists. Such unknowable (for the Spanish) elements indicated a defiance to colonial control, making them a source of anxiety. Likewise, cartographer Manuel García-Baquero found that prolonged sandstorms provoked not only physical effects on humans but also psychological ones.[57]

These trickster winds affected nonhumans too. Benítez noted how the winds deceived the camels and how winds of different temperatures and humidities affected their behavior:

> When the camel has not drunk for a day, and thirst starts to tire him, he starts walking as fast as possible in the direction that the wind blows, attracted by its coolness, by the belief that he is walking to the point where he will find water to drink, and since this means that he would get separated from the caravan, and separation from the caravan means losing oneself beyond salvation, the guide will not stop recommending to us that we follow him, and that we do all we can to oblige our camels to follow his camel. . . . This danger stopped as soon as the dry and hot desert wind blew, because then the camels remained obedient to the guide without thinking of the water that for a moment they thought they would find when they felt a gust of cool wind, uncommon in the Sahara.[58]

Deceived camels aside, not all colonial writings made a terra nullius of the Sahara. Some writers offer nuance. (French) Antoine de Saint-Exupéry said of the desert around the Spanish fort at Cape Juby (just north of Western Sahara today), "If at first it is merely emptiness and silence, that is because it does not open itself to transient lovers."[59] Cervera and his team, even though they traveled at the height of summer, noted that most dunes were limited to the coast. Inland, there were abundant vegetation, trees (rubber, carob, and holm oaks), and lakes of several kilometers in diameter.[60] Emilio Guinea López's writings also contradict the terra nullius narratives constructed by the majority of his scientist compatriots. Guinea López, a renowned Spanish botanist, wrote three books about the botany of Spanish Sahara based on a 1943 expedition undertaken with Hernandez Pacheco and Vidal Box, sponsored by the Institute of Political Studies. The expedition had aimed to increase knowledge of Spanish Sahara's natural resources. Guinea López's book on the expedition itself was written with a public (nonexpert) Spanish audience in mind.[61] Making ample use of literary flourishes, Guinea López looks upon "their" desert with romantic affection and absolute enthusiasm. He admits, in his prologue, that, ahead of his expedition, the desert represented an empty hole (*vacío*) for him, that his attitude toward the desert was one of hostility, and that he expected to find little that would interest him with respect to botany.[62] Yet, upon embarking on the expedition, Guinea López found himself immediately "bewitched" by the desert,[63] impressed by its "strength," its abundant "character," and the "new sensations" it offers for Europeans.[64] The plants that he encounters seem to show courage and ingenuity in their climate-motivated adaptations. Thus, the desert becomes a creative space, rather than a morbid emptiness, in his writing. Guinea López is liberal with exclamation marks in his descriptions of the desert and goes as far as to say that the days and nights he spent there will be the source of his happiest memories when he becomes an old man.[65] He ends his book with the poignant "optimistic hope of return."[66]

For Guinea López, the Saharan wind is the "owner" and "creator" of Western Sahara's desertscapes,[67] and it works "with great artistry" in sculpting the land,[68] moving dunes into amusing croissants and moustaches[69] and creating filigree-like latticework on boulders.[70] The wind "sweeps,"[71] "beats,"[72] and "scratches"[73] the land, frequently violently, sometimes "playfully,"[74] and occasionally even "erotically."[75] For Guinea López, the winds are not terrifyingly barbaric, but they are captivatingly beautiful. Unsurprisingly, the botanist pays much attention to winds' effect on vegetation, and, like his compatriots, writes at length on the impact of winds, particularly the *īrīfī*, on the daily activities of his expedition members. Nevertheless, unlike his compatriots, he draws attention to the opportunities created by the wind's modification of their behavior, as is evident in the following quotation: "The incessant wind that now blows in our faces (we are walking towards the North) obliges us to march curved over, with our heads down, which is useful for observing the ground."[76] Enrique D'Almonte is also sensitive to opportunities. Rather than seeing the *gallāba* as little more than barren, windblown rocky outcrops, he notes their navigational possibilities as landmarks that could be seen from dozens of kilometers away as well as their political potential as boundary markers.[77]

Nuance likewise characterized Spanish colonial writers' attitudes to Indigenous epistemologies. As Sampedro Vizcaya highlights, colonial environmental sciences have been dependent on unacknowledged local knowledge.[78] In Spanish Sahara's case, Saharawi guides, as they were called, provided vital knowledge that informed and shaped colonial research publications and kept Spanish researchers alive and well during fieldwork. Spanish attitudes toward Saharawi knowledge ranged from annoyance to outright invisibilization to casual sidelining to exoticization and admiration. At the annoyed end of the scale, we find Julio Cervera. Cervera was scathing, condescending, and racist in his observations of the Saharawis. He shrugs off their cautions about the winds' dangers and is irritated when the Saharawis wish to pause the journey upon the arrival of the *simoon* (*īrīfī*): "The *simoon* wind, accompanied by a sandstorm, worked like their assistant. They did not want to move forward, using as an excuse the loss of five camels and the danger of marching through the desert during a storm."[79] To ignore the Saharawis' superior knowledge of how to deal with the local climate showed typical colonial arrogance. Yet, as José Antonio Rodriguez Esteban has highlighted, Spanish explorers, military personnel, and scientists were completely defenseless in sandstorms and relied time and time again on Saharawi guides to save them.[80]

Other authors give subtle nods to their Saharawi guides. Hidalgo de Cisneros, for example, was responsible for supervising postal operations and ensuring the safety of regular French transatlantic services and cartographic reconnaissance flights, for which—he mentions in his writings—he relied on a "Moorish" assistant.[81] Likewise, Congost Tor was able to complete his ornithological expedition thanks to a "friendly" Saharawi guide named in the resulting publication only as Chacote.[82] Narbón claims, in the blurb to his book, to have spent prolonged periods in the open desert and "navigated [it] as if it were an ocean of dry land, using instruments that belong to sea navigation," and yet other writers cited here claim their compasses, pedometers, and other such instruments could last but a few days in the desert.[83] Inside Narbón's book, there are several Saharawis depicted in photographs of his expeditions, and his affectionate and long friendships with some, such as Brahim, are elaborated in the text. One might assume—but Narbón does not confirm—that his successful navigations were facilitated by his Saharawi friends.

In colonial research publications, the passive voice often indicates hidden Indigenous knowledge. Emilio Guinea López "is told" by an unspecified third person plural, whom we can reasonably assume to be Saharawi, how to survive if caught out by the *īrīfī*.[84] However, he compares the sophisticated knowledge of Saharawi nomads of desert flora with the "ignorance" of the Spanish rural peasantry. Guinea López says that Saharawis' knowledge of vegetation was a "pronounced and unforeseen resource" for his work.[85]

One can also trace the tracks of Saharawi participation in the vitally important Spanish 1:500,000 scale map of Spanish Sahara. The mapping project, led by Manuel Lombardio Vicente of Geographical Services, began in 1943, took seven years to complete, and relied heavily on Saharawi knowledge and labor for its production. Maps can be exploitative things. This one was used for all developments in Spanish Sahara, including the localizing of minerals and hydrocarbons, and, as Rodriguez Esteban notes, is still a key reference even today.[86] The Saharawi contribution is immediately clear when one glances at the map even for the first time: all three thousand toponyms are transcriptions of the Saharawi words for each place. The use of *Saharawi* terms, with *Saharawi* pronunciation, was vital to the safety of Spaniards. Because the colonizers relied on Saharawis for navigation, especially, as Rodriguez Esteban points out, when it came to locating life-sustaining wells, it was imperative that Spaniards could communicate effectively with Saharawis.[87] To replace Saharawi words for landmarks with new Spanish ones would not just have been counterproductive but also dangerous. Nomads' classifications of, and terms for, topographs were therefore carefully transliterated for the map.

While Lombardeo Vicente failed to refer to Saharawi participation in forming the map during the various conference presentations he delivered on its production,[88] geographer Angel Domenech Lafuente formally acknowledged the contributions of three named Saharawis (Chej uld Maaif uld Borhim, Reguibi; the "mudarris" [teacher] Si Hosain b. Mohammed ben Ali, Asargui; and the "fakih" [jurist] from the Marksmen Squad, Si Maimún bel Hach Ahmed).[89] Another member of the Spanish Army's topographical team, Ángel Flores Morales gives a more detailed account of the phenomenal knowledge of his Saharawi guides. Writing in 1949, he states the following:

> They have a very developed sense of orientation. Their memory conserves the most difficult panoramas, even ones they've only seen once. . . . A completely uneducated nomad, if asked by the head of the expedition, will draw a sketch with a finger or a stick in the sand that shows the point of interest. The nomad has a sense of topography, since for him direction is a question of life or death. He is a great recogniser of tracks. He can identify an individual from any given tribe by a footprint in the earth, and he does so with as much reliability as a policeman from a civilised country identifying a criminal by his fingerprints. When he loses a camel, he doesn't mess around, he goes straight to the nearest well and there he searches among the hundreds of footprints that other dromedaries left while drinking, coming from completely different places, and he follows the track of his animal, which he recognises immediately . . . until at last he finds the camel, and he returns to his *jaima*. . . . It's not out of the ordinary, when travelling in a caravan, to see a guide laying face-down on the ground and sniffing thoroughly. His sense of smell won't let him down: he will say within a moment if there is water nearby in the «*daias*» formed by the rain or if there's none left. It's fantastical, but it's the reality. If they see a flock of emigrant birds, they carefully observe them, and they pay attention to

where they come from and which direction they are flying in. Years later, when passing by the same spot, they'll say "There's water there, because once I saw some birds heading in that direction."[90]

Despite the earnest admiration of Flores Morales and others, Saharawi contributions to colonial science (including saving the lives of the Spanish scientists upon the arrival of the *īrīfi*) remains in the shadows. Saharawi assistance was rarely fully acknowledged. Meanwhile, winds in colonial (pseudo) science writing reveal aeolian anxieties: wind was a barrier to the colonial project. But Spanish writers also made Saharan winds their tools, using them to discursively construct a windblown terra nullius, as well as a windblown Indigenous community, allegedly in need of colonial modernization. These windblown discourses would later inform energy infrastructure in 1960s Spanish Sahara.

Electrical Infrastructure Development in Spanish Sahara

Mahmoud Alina was born into the sweet nomadic life in Tiris, a renowned part of the *bādīa* nicknamed *bride of poets* for its inspiring beauty. In that life, when the wind moved, it told Mahmoud where to take his camels, his goats, and his sheep for water and pasture. He could also read the air's movements to predict dangerous winds and storms, and thereby avoid catastrophe. He used to prepare for the *īrīfi*'s arrival by filling goat skin bags with water and storing them beneath blankets in the *ḫaīma*, to keep them from the flames of the wind. All travel plans would have to be halted when the *īrīfi* blew. Following such heat, Mahmoud would hope for a cool, smooth *Sāḥlīya*, a wind that would "open the land and refresh its pores."[91] Depending on the season, the *Sāḥlīya* might bring rain, or, in the summer, fresh, mild weather, perfect for traveling. Best of all was when the *Sāḥlīya*, or another good wind such as the *Gblīya*, brought *Garūāḥ* wind. *Garūāḥ* means a relaxing time for everyone: animals, humans, and plants. "The *Sāḥlīya* is smiling now," say the nomads in the presence of *Garūāḥ*.[92] Saharawis also make a verb of this wind (for example, the second person feminine present conjugation *tatagaruḥī*, or second person feminine present tense *ātagarwaḥtī*).[93] A friend defines the meanings of this windy verb, showing how it represents time, place, emotions, and senses all in one:

1. To be in a place of fresh air.
2. To have peace in a peaceful place. For the Saharawi, the verb is usually used with reference to time spent in the *bādīa*, but not exclusively so.[94]

However, hot Saharan easterly winds, while able to cure some human ailments, burn plant life.[95] In the 1960s and early 1970s, such winds provoked droughts that decimated Saharawi camels and goats. Nomads were used to battling through such hard times. The problem was that colonialism curtailed their traditional adaptation strategies.[96] The droughts therefore forced several thousand Saharawis to the cities, often temporarily,[97] only as long as was necessary to survive, to look for employment opportunities in the phosphate mine, in local businesses, as merchants, as drivers, or in the army. Mahmoud and his family moved to El Aaiún permanently in these times of drought. He became a butcher on the payroll of the Spanish civil service. Today, he still lives in El Aaiún and still has some of his Spanish papers. Now, he is connected to the grid. Back in the Spanish era, fire was his original energy source, and later gas cylinders for heating food once he settled in the city. Unlike his Spanish colleagues, he did not have networked electricity at home.

Spain developed energy infrastructure in 1960s Spanish Sahara with the primary purpose of servicing the oil and phosphates industries and the residential needs of their respective Spanish (or other white) employees. However, electrical installations were in a constant state of insufficiency. Despite the pleas of businesses,[98] civil servants, and residents[99] to the governor of the Sahara, and in turn, his pleas to Madrid for more budget, infrastructure did not grow at anywhere near the pace of demand.

Over the 1960s, authorities repeatedly deemed El Aaiún's electricity supply to be grossly insufficient and unsafe.[100] This had a critical impact on the city's drinking water supply, which needed electricity to work reliably.[101] Equally problematic were the hospital's supply issues. In November 1968, an electrical storm provoked discharges from high tension power lines, which in turn, caused a power cut in the hospital. The blackout persisted for most of the day, with an obviously negative effect on staff and patients. In response, the hospital demanded an independent generator.[102] By 1970, the government forked out for an expansion of the city's power station, but the bright lights of shop fronts on the high streets were still out,[103] and the lack of supply frequently made television broadcasts impossible.[104] In 1971, El Aaiún's mayor complained that a lack of street lighting was causing hooliganism in some parts of the city.[105]

The situation in other Spanish Saharan cities mirrored that of El Aaiún. A civil servant responsible for public works described the state of Villa Cisneros's electrical supply as "truly distressing" in October 1966.[106] A few months later, a desperate governor-general sent a pleading letter to the director general of African Posts and Provinces (based in Madrid) listing several public and military buildings, as well as entire residential areas in Villa Cisneros, with no connection to the grid.[107] The electricity infrastructure in Smara and La Guera also remained incapable of meeting the demands of local industry and residents.[108]

Access to electricity was more prevalent in the most affluent (Spanish-dominated) suburbs. Saharawis felt they were last in the queue to benefit from extensions to the grid. Saharawis usually made use of charcoal, candles, and lamps for their domestic energy needs. Zrug, a resident today of Moroccan-occupied Western Sahara, recalls, "Spain did develop electrical facilities, but these were only for lighting some public spaces, for government buildings and for the homes of Spanish officials."[109] Mahjoub, a fisherman born in Spanish Sahara, however, reports that some Saharawis did benefit from electricity during the Spanish era. He recalls families that had washing machines, for example.[110] In El Aaiún in 1974, 102 Saharawis signed the following petition to the governor-general: "Those on the following page, who are all of legal age, and are traders and residents of this City, with homes or businesses in the Catóñico cemetery neighbourhood, with due respect and consideration write to you to dare to explain: seeing ourselves deprived of the necessary electrical light connection, which the rest of the population benefits from, as is the case for the sewerage network, we request that you give the orders necessary for us to benefit from both networks, especially the electric current."[111]

Saharawis in Smara made similar demands to the governor-general.[112] Through such petitions, electrical infrastructure reveals itself as the battleground between colonizer and colonized. Exclusion from resource infrastructure such as energy grids or sewerage systems has a long history in colonial contexts as a tool for domination and racial segregation.[113] Through their petitions, Saharawis showed their resistance to such oppression and made electrical infrastructure a battleground between colonizer and colonized. If Saharawi residents complained via petitions of their lack of access to electricity, others that worked on the supply side were equally prepared to fight for their own rights. Saharawis that worked as electricians, technicians, and guards in the power stations of Spanish Sahara requested lodgings and demanded pay rises.[114] Efforts to gain political authority are often expressed in the form of electrical infrastructure projects.[115] Spanish infrastructure enacted a racialized authority over Saharawis, but the latter simultaneously made the infrastructure a site of resistance.

While networked electrical infrastructure mediated a colonial, racist politics in the segregation of its services, it also colonized at the level of aesthetics. It is possible to read energy (infrastructure) as (an) aesthetic entity/ies. By doing so, I argue that energy infrastructure served as an aesthetic vehicle for colonialism's cause. I make two key points: first, energy infrastructure served as the aesthetic representation of colonialism's triumph over the disorderly Saharan winds and windblown desertscapes, and second, Saharawis recall changes in energy source in aesthetic terms. In its self-presentation and its aesthetics, energy infrastructure relied on, and transmitted, the cultural understandings of wind developed in wider Spanish colonial writings. In turn, the energy system's impact produced aesthetic changes in the lives of Saharawis. Following Nadera Shalhoub-Kevorkian in her work on the Israeli "occupation of the senses" enacted on the Palestinians of east Jerusalem, I conceptualize the development of energy infrastructure in Spanish Sahara as a colonization of the senses.[116]

Spanish Sahara's energy supply depended on imports from Spain. Atlas S.A. Fuels and Lubricants (the company later changed its name to Atlas S.A. Fuels and Lubricants Sahara), headquartered in Ceuta, supplied Spanish Sahara, including its power stations, with gas oil. The Aaiún Playa factory, which opened in late 1961, served as a warehouse and reception for imported fuel. As its name suggests, the factory was located on the coast near El Aaiún city. Protruding from the factory seaward were two kilometers of pipe, which allowed the oil tankers to unload their cargo directly into the factory's storage facilities. In a company-authored dossier dated 1961, Atlas boasted of its technological prowess with regard to this underwater plumbing: "For the installation of this submarine piping, the first of its kind in Spain, it was necessary to overcome serious difficulties caused by the strong marine currents and ruling winds of that zone."[117] The tankers brought gas oil from the Spanish Petroleum Company (CEPSA) Santa Cruz refinery in the Canary Islands.

In the city, the El Aaiún factory (opened in 1960) had three tanks, each of which held up to 150,000 liters of gas, oil, or petrol and had "filters that allow[ed] perfect filtering . . . preventing the entrance of any sand."[118] Lorries transported Atlas's petroleum products around Spanish Sahara. Atlas explained, "To load the lorry-casks and also the cans, the Factory has a wide store of useful stock with jointed pipes that allow, additionally, the avoidance of all contact between the fuel and the sandy environment."[119]

Atlas S.A.'s descriptions of its own installations repeat the tired, colonial tropes of human technological prowess triumphing over hostile nature, colonizer over colonized, and modernity over the backward. The winds, marine currents, and whirling desert sands accentuate the hostility of nature and therefore the proportionately increased genius of Atlas's technological and engineering solutions. Sketches of the two factory sites, held in the aforementioned 1961 dossier, add further insights into Spanish colonial energy discourses.[120] They draw on the colonial imagination's desert, which is characterized by blandness, lack, and waste. Both Atlas sites are sketched on a flat, empty desert—a terra nullius devoid of aesthetic charm. The land is rocky, and the sole flora are spikey cacti and another plant with bladelike leaves. The plants' sharp extremities conjure a touchscape of hostility. The perspective of the sketch ensures that the Atlas site, in both cases, occupies most of the paper. The lines are harmoniously straight and crisp; the outhouses and even the lorries are drawn at perfect right angles to the border walls of the sites. It is implied that Atlas, and indeed the Spanish colonial system, bring order and logic to the wayward desert. The sketches suggest that Spanish energy infrastructure has overcome the hardships of heat, dust, sun, and wind. The desert has submitted to development.

Photographs of the sites are aesthetically futuristic. Aaiún Playa's three tanks are enormous cylinders, with narrow staircases spiraling elegantly up their sides. Their white color and hard, smooth surfaces suggest refuges of cleanliness and purity in contrast to Cisnero de Hidalgo's aforementioned nightmare of windblown aerosols. A photograph of the Aaiún city factory shows three tanks set on stilted platforms, as if they were rockets waiting to be launched to space, the final colonial frontier. On the sketches, these tanks have the word *futuro* drawn onto their sides like badges, suggestive of groundbreaking technology and ultramodernity.[121]

A great theoretical advantage of thinking through the wind is that it forces one to pay attention to the senses—and therefore to aesthetics. One might imagine how the towering Atlas fuel tanks painted white would have made you squint and seek to cover your eyes in the Saharan sunlight, as if designed to transmit a celestial power over all who came near them. Their enormity, and the fences all around them, constituted spatial domination. Energy infrastructure transformed Saharawis' daily sensory experiences in many such ways. Atlas credited itself with having "introduced the use of butane gas to the nomad population of El Aaiún."[122] As well as conquering the desert's hostile climatic conditions to power Spanish Sahara's electrical infrastructure, Atlas brought added value by selling the natives gas cylinders. Nguia, an environmental activist and neighbor of Mahmoud, remembers when gas came on the market. A colleague of mine asked her when she first had access to such energy in the home. She responded as follows: "I have been using it since 1970 in my family home. I was six years old. I used to see my family cooking under coal, and in 1970, the gas came. Cooking with gas has a different taste to coal."[123]

Given the brevity of her response, it is interesting that Nguia remarked upon—without a request for her to do so—the sensory aspect of the change in primary energy source. Volunteering this information arguably reveals the importance, to her, of the aesthetics of flavor attached to traditional Saharawi energy sources. Her lingering memory of the colonizer's new energy is implicitly one of loss. When butane gas entered her home, smoky sensual pleasures drifted away. What other aesthetic changes would the introduction of Spanish colonial energy infrastructure have brought? What aesthetic opportunities appear and disappear when a coal or deadwood fire is replaced with a gas cylinder or a mains electricity-powered oven and lamp? And what of the aesthetics of pylons, cables, and transmission lines? To consider such questions, let us again think of Nguia and her family cooking bread under coals and frying meat in a pan that spits oil above the flames. Such a form of cooking, in which the chef digs a hole in the sand and buries the dough before lighting a fire on top, is necessarily outdoors. The outdoor hearth drew extended family and neighbors together, allowing for social opportunities, exchange of stories and poetry, living in the weather-world, and daily intermingling with people, the wind, and the world. One might imagine the warmth and glow of the fire, the soft feel of moving air on your skin and sand underfoot, and the voices of the poet, the storyteller, and the playing children. Butane-powered gas stoves, and later networked electricity, pushed the family inside into a contained, controlled space that kept the desert elements out. Perhaps people of all communities can appreciate the aesthetic difference of the flickering glow of a campfire below the stars compared to the stark light of electric bulbs and unromantic odor of a gas stove at combustion. On the introduction of networked energy in Europe, Wolfgang Schivelbusch comments, "When the household lost its hearth fire, it lost what since time immemorial had been the focus of its life . . . fire had always remained clearly and physically recognisable as not merely a product but also the soul of the house. . . . People no longer lost themselves in dreams of the primeval fire; if anything, they were thinking of the gas bill."[124]

Let us imagine, too, the politico-aesthetic implications of electrical infrastructure networks where, previously, only independent energy sources had existed. Centralized power means dependence and sedentariness. Before, fire meant independence, flexibility, and mobility. Centralized electrical infrastructure itself represents aesthetically the loss of autonomy of producing one's own light, heat, and cooking source. It reaches, from a centralized source, with tentacle-like cables, into each private home.

Nguia's briefly volunteered remarks on flavor, then, reveal that energy systems are aesthetic projects as much as political, economic, or technical ones.

Focusing on the aesthetics of energy infrastructure from a Spanish colonial perspective offers further support for this conclusion. In the Western world of the late nineteenth and early twentieth centuries, electricity was a symbol of progress and modernism.[125] This was also true for colonists in European colonies.[126] Although Spain rolled out energy infrastructure primarily to favor natural resource exploitation, energy networks also served to transmit colonialism's self-justification message of bringing light to darkness, metaphorically and literally. The dazzling lights of electric bulbs in public buildings, private homes, and the streets were an aesthetic reminder of Spain's civilizing mission in the Sahara. The extended networked infrastructure visually indicated state power.

But storms frequently assaulted this symbol of state power. As such, controlling the weather was a constant concern for the Spanish government. Desert wind, rain, and sand constituted an ongoing threat to colonial infrastructure. Floods and high winds, in particular, repeatedly damaged infrastructure.[127] Civil servants duly reported on these wind-induced disruptions in telegrams and letters back to Madrid and, from the late 1960s, via standardized and regular meteorological reports. In Spanish Sahara, police manned meteorological stations in Cape Boujdour, Boujdour town, Mahbes, Smara, Echdeiria, Tifariti, Hagunia, Guelta, and Hausa. They sent uniform daily reports to the military air service, which commented on wind direction, speed, and strength as well as temperature, visibility, and levels of rain. A descriptive column on the generic meteorological report template allowed for additional notes, where police occasionally named winds, such as the "siroco."[128] For the most part, though, the words used for wind were standardized, had only a practical use, and borrowed from the precise terminology of seafaring (that is, winds were named by their direction). This is hardly surprising since twentieth-century European colonial practices of observation, when it comes to weather and climate, were a continuation of the maritime tradition of previous centuries in which close monitoring of weather patterns was a life-or-death matter.[129] Observations on the wind were, through these stations and processes, institutionalized and put at the service of the imperial project.

Controlling the desert atmosphere was a high-stakes business not just for colonial aspirations in general terms but also specifically for energy production. Wind was a force that threatened future economic prosperity. The oil company Esso Standard Española S.A. lobbied the governor-general to undertake measures to stabilize what are perhaps the world's best-known aeolian features: dunes. In 1972, in what he described as his "continuous search of information about the subject of controlling the sand of the desert," Gerardo Alvarez de Benito, an Esso employee, sent a file of research to the governor-general.[130] It contained several academic studies of the use of cutback bitumen to stabilize sand in other locations in Africa where hot winds were creating sand dunes that prevented (neo)colonial exploitation of mines. Alvarez urged the governor-general to heed the research on how to construct roads that could service the oil industry in sandy Spanish Sahara.[131] Just as the desert's winds represented the chaotic in the Spanish colonial mind's eye, energy infrastructure was a colonial means to create controlled, contained spaces and thereby tame the winds.

But the colonized can equally invoke the wind in imaginaries of resistance. "May the Red Wind come after you" (يعمل الريح الحمرا من دونك) is a common Saharawi curse for someone that you wish would disappear or die.[132] A Red Wind is red because of the sand it carries. It is a great sandstorm from the east, seemingly infused with all the heat of the vast deserts of Algeria or Mali, ready to engulf and bury everything that lies ahead.[133] As Nguia confirms, the Red Wind buried houses, cars, and camps in Spanish Sahara in 1945 and 1949.[134] Mahmoud describes it as "a wind that brings red sand and dust and it hides the sky, so the sky becomes a red flame. You can't see clouds or sun or anything."[135] Saharawis can invoke the wind's power in such speech acts, blowing away their enemies.

Saharawis invoked winds of resistance in their writings during the Spanish colonial period. Although the Saharawi tradition is oral, young students of Spanish-run secondary schools were publishing prose and poetry in the 1960s and 1970s in, for example, school magazines. One of the most famous of these was *Irifi*, the student magazine of General Alonso High School, El Aaiún, published between 1967 and 1972. According to scholar (and Friendship Generation poet) Larosi Haidar, works in the magazine attempted to evoke what it means to be Saharawi (*saharauidad*).[136] For Haidar, the name of the magazine was a primordial part of this wider expression of Saharawi identity, and he notes how this namesake—the *īrīfi* wind—often appears as a central focus in the students' work. Furthermore, and as Haidar argues, Saharawi students' prose and poetry in *Irifi* reflected, through metaphors, their political awakening.[137] The hot easterly wind, which the Spanish colonizers considered as destructive and terrifying, was the perfect metaphor for the young students' eagerness to resist and fight for independence.

Nationwide, organized movements to rid the Sahara of its Spanish colonizers emerged in the late 1960s and picked up pace in the early 1970s.[138] Mahmoud Alina was a friend of Mohammed Bassiri, who would become the first martyr of the Saharawi pro-independence movement.[139] Bassiri was born in Tan Tan in 1942. Tan Tan today lies within the internationally recognized borders of Morocco. In 1942, it was part of the Spanish protectorate. Before Spanish colonization, it was in the northernmost part of lands roamed by Saharawi nomads—lands outside the rule of the Moroccan sultanate and Mauritanian emirates, lands that stretched to the Draa river in the south of modern-day Morocco. The year 1958 saw the end of the Ifni War, when Spain handed the Ifni region to Morocco.[140] Bassiri's hometown, Tan Tan, thus became internationally recognized as Moroccan, as it lay within this region. For Saharawis, who were subject to terrible human rights abuses under their new Moroccan rulers, this was a Spanish betrayal and a first step in Morocco's grand expansionist plan—the newly independent Moroccan state envisaged a Greater Morocco, which would recover an imaginary historic empire and encompass all Western Sahara and Mauritania, as well as western Algeria and Mali. These expansionist plans saw Morocco laying claim to Mauritania until as late as 1969 and, in 1963, invading parts of western Algeria. The latter provoked the War of Sands, which raged on until the Organisation of African Unity brokered a precarious cease-fire between Algeria and Morocco the following year. Saharawis therefore had well-justified fears, from as early as 1958, that Spanish Sahara would be handed to land-hungry Morocco. This partly fueled the great popularity of emerging nationalist movements, including that which Bassiri would lead.[141]

Bassiri studied in Morocco, Egypt, and Syria, where he became well-versed in the ideological currents of the day, including Pan-Arabism, third world solidarity, and socialism. This shaped his ideas about Western Sahara's future. Back in Spanish Sahara, Bassiri lived and worked in Smara, where he opened an Arabic school, which became a hub for discussing nationalist ideas. In 1969, Bassiri and his colleagues established the anti-colonialist party the Vanguard Movement of Saguia el Hamra and Rio de Oro (named after the two regions of Spanish Sahara), or Liberation Movement for short. It (clandestinely) recruited thousands of Saharawi members very quickly, including Mahmoud Alina, who, with a civil service job, was able to contribute to the funds of the party.

Bassiri and his comrades organized a huge, peaceful demonstration in Zemla Square, El Aaiún in 1970. The Spanish Legion opened fire on the thousands of Saharawis protesting for independence, killing at least eleven, and arrested Bassiri. Most likely, police executed him shortly afterward. This was a turning point in the Saharawi independence struggle. Victory would not be won by peaceful means. Voices in favor of an armed independence revolution were allegedly heard "within an hour" of Zemla's bloody end.[142] A young man from Tan Tan, El Wali Mustapha Sayed, played a leading role in advocating this path. He and fellow university students, Saharawi soldiers serving in the colonial army, and Saharawi civil servants formally founded the Popular Front for the Liberation of Saguia el-Hamra and Río de Oro (Polisario Front) in 1973. Its guerrilla army sought to hit Spain economically through acts of sabotage. One of Polisario's key successes in this ambit was its destruction of the phosphates conveyor belt, which transported phosphates from the inland mine to the port.

After dark on October 19, 1974, two pairs of Saharawi electricians and electrical technicians, all employees at the Fosbucraa mine, led teams of four Saharawi pro-independence activists to stations 7 and 8 of the mine's conveyor belt (over one hundred kilometers long and constructed at a cost of $72 million). The mission of the electrical teams was to render useless the TT-40s, devices used to detect the location of faults on the belt, and to destroy all electrical installations at station 7. They also pulled off the wooden panels that covered each station's electrical cables, doused them in gasoline, and set them alight, destroying three hundred meters of belt.[143] The Fosbucraa attack was one of Polisario's principal acts of sabotage. Spanish police caught one of the Saharawi electricians responsible for the attack and tortured him for three days, until he gave up the names of his co-conspirators. They too were tortured into signing confessions. Energy infrastructure became a site of violence, a characteristic that would only intensify with the Moroccan invasion precisely one year later.

Conclusion

Emilio Bonelli's first official colonial expedition to Spanish Sahara came shortly after the so-called Age of Sail (c. 1600–1850). When writing up his findings, he drew on seafarers' metaphors of wind, boat, and storm to communicate his vision for economic exploitation as political progress in Spanish Sahara:

> Society cannot be governed as easily as a ship that overcomes the setbacks of the wind, and, therefore, once the valves of progress are open, the impetus of revolution that operates in a social body corrupted by barbarism can only be contained by anticipating the necessary elements for counteracting the effects of the storm via a considered political plan. The latter should be completely subject to the conditions of the country and its inhabitants and in harmony with the trade and equilibrium of all the forces of production.[144]

Links between the shipwrecked's narratives of the Sahara and Spanish colonists' writings go beyond mere maritime metaphors. They also share the aeolian anxieties provoked by Saharan winds and the windblown. Winds highlighted the limitations of the Spanish colonial project. Wind broke colonizers' technologies, infuriated them, and threatened to bury them alive. The Spanish could not conquer the winds, and when they wanted to move through the desert, they relied on Saharawi guides to do so. Nevertheless, much like the shipwrecked, the Spanish, for the most part, made Saharawis' knowledge, skills, and contributions to colonial science invisible. Colonists used wind and windblown imaginaries to place Saharawis on the inferior side of the culture/nature and civilized/barbaric binaries. I should highlight, though, that the Spanish's lexical spectrum of orientalist discourse on Western Sahara had evolved from that used by the shipwrecked. The tone in Spanish colonial science writing is one of benevolent condescension. This constitutes a move away from the shipwrecked's more crude discourses of Saharawi savages and monsters.

As was the case for the shipwrecked, the anxiety-inducing wild winds served to reinforce colonizing Spain's moral conception of itself. Colonial writings orientalized the desert winds. The winds were terrifyingly barbaric, thereby reinforcing the moral, civil respectability of the colonizer and making the Western Saharan weather-world the Other (the product of a colonial process of *Othering*, in which the colonized Other is marked as lesser and marginalized), ripe for conquest. Wind imaginaries, developed during expeditions that would facilitate resource exploitation, were colonial instruments for justifying and consolidating authority over the desert terra nullius.

Wind and aeolian desertscapes, descendants of sailors' earlier imaginaries, underpinned Spanish colonial discourses in Spanish Sahara. Energy infrastructure, like other built infrastructure, extended in Spanish Sahara was a means of controlling and colonizing the wild and barbaric windswept desertscapes and of making them useful according to capitalist logics. In its management, energy infrastructure enacted racial segregation, and its aesthetic form, as well as the daily aesthetic experiences that it transmitted, also furthered colonialism. The Spanish energy system was underpinned by, and in turn reproduced, wind-infused colonial logics.

But, riddled with aeolian anxiety, the colonizers wrote of the Saharan winds as if they were powerful agents. And powerful agents they were. The Saharan winds dictated the form of the first energy infrastructure developed in Spanish Sahara (the windmill running a dynamo at the fort), they undermined and destroyed later electricity networks, and they had to be accommodated with regard to the logistics of importing fuel to generate electricity. Throughout the colonial period, Spanish scientists dared not set foot in the desert interior without an Indigenous guide for fear of live burial at the hands of the Red Wind. For all the pompous language concerning infrastructural victories over the desert, no one could navigate the Saharan winds like the Saharawi nomads, and, one could argue, no one ever will.

Part Two
Occupation

CHAPTER 3

The Tamed Winds

Siemens and Settler Colonialism in Moroccan-Occupied Western Sahara

---

Saleh[1] tells me to take photographs. We are still within Morocco's modern internationally recognized borders, but these are ancient Saharawi lands. Spain handed them to Morocco in 1958 following the Ifni war. Today, the Tarfaya Strip, home to one of Africa's largest wind farms, still has a sizeable Saharawi population. Saleh decelerates and winds down the window with his free hand so that the pictures will not be clouded by the dusty glass. The wind barges in, bringing sand. I squint and hold up my phone. Outside, a squadron of Siemens turbines are triumphantly staked into the ground.

This chapter is the first of two that discuss the interrelation between the human rights black hole and the Moroccan energoregime in occupied Western Sahara. I focus on how wind energy is presented to audiences external to Western Sahara and Morocco. First, I explore how Morocco uses wind and solar factories in occupied Western Sahara to do its diplomatic work abroad, exchanging energy for implicit acceptance of its illegal occupation and greenwashing its crimes. Second, I focus on how the company Siemens, behind several wind energy developments in occupied Western Sahara, uses colonial discourses to present its activities in promotional materials. I argue that Siemens directly furthers Morocco's settler-colonial project in occupied Western Sahara. The company builds on entrenched imaginaries—forged and furthered by European and American shipwreck survivors and Spanish colonialists—of Western Sahara as a wind-ravaged wasteland. I contend that

Siemens uses such wind imaginaries to justify its legally and morally questionable activities. In particular, Siemens relies on colonial wind imaginaries to construct Western Sahara as terra nullius. Here, Siemens's discourse is a synecdoche—one grain of sand standing in for a desert—for corporate energy companies' colonial discourse more widely. Today's energy corporation wind imaginaries are shaped by the colonizers that came before. These imaginaries continue to reinforce modern-day colonial projects.

I use thick description, particularly regarding my entry into Western Sahara in this chapter and when referring to violence suffered by Saharawi protestors in chapter 4. I do so to give the reader a clearer sense of the climate of oppression in occupied Western Sahara. Such an insight is necessary to understand the legal aspects surrounding wind energy developments. For such activities to be legal, energy companies active in occupied Western Sahara would need Indigenous Saharawis to consent to their activities.[2] Previous court cases indicate that, specifically, Polisario (the only UN-recognized representatives of the Saharawi people) consent would be needed.[3] Several energy companies claim that their activities are legal because they have consulted with the local population in occupied Western Sahara. They neglect to distinguish between the Moroccan settler population and the Indigenous Saharawi population. They also neglect the issue of political freedom in Western Sahara. The Washington-based NGO Freedom House regularly lists Moroccan-occupied Western Sahara in the small "Worst of the Worst" section of its annual ranking of global political freedoms by country. Saharawis are simply not allowed to express their opinions. The purpose of my lengthy descriptive sections, then, is to show a context in which to ask the reader a rhetorical question: in this climate of oppression, could an energy company—Siemens, for example—ever obtain freely given, uncoerced Saharawi consent?

When Saharawis protesting against illegal natural resource exploitation are beaten, locked up, and even disfigured, no one reads about it. The UN presence in Western Sahara is the only modern peacekeeping mission without a mandate to monitor human rights. If one wants to hear Saharawis' stories of occupation firsthand and write these stories up so that others elsewhere can later read them, one has to get into Western Sahara unnoticed, bypassing the Moroccan state's extensive secret police and civilian informer network. Crossing the border gives a traveler a first taste of the oppressive context in which the energoregime is built.

Glimpsing a Wind Factory on the Way to Western Sahara

It is summer 2014. The wind flicks sand in my face and makes a kite of my long blue scarf. Abdelhay[4] catches it. He then rewraps it around my head in grand folds until it is a Saharawi turban again. He tells me to put the sunglasses on for the photographs. I am now covered head to toe—my blond hair and sunburn well hidden. Abdelhay and Saleh have dressed me in combat trousers and a baggy camouflage shirt, with a small Spanish flag embroidered on the upper arm. I look just like a Saharawi man. Men cover their faces to keep the sand off, or to show modesty around women outside of their family. Thankfully, such customs make it easy for my friends to change my gendered and raced appearance. We all make the "*V* for Victory" sign and laugh with joy—we have almost made it. Saleh zaps the images to our mutual friend in London via WhatsApp. We all agree that Naseem[5] will be delighted.

Naseem directed the first stage of this trip from his flat in Camden, North London. Abdelhay, Saleh, and I were in Tarfaya just before dawn. It is a small Moroccan town famous for being a temporary home to Antoine de Saint Exupéry, author of *The Little Prince* and lover of the Sahara Desert. I was on foot, alone, and just descended from the coach that departed Marrakesh the previous afternoon. Abdelhay and Saleh were in a 4x4. All of us were on the phone with Naseem in London, who held Nokias with new pay-as-you-go SIM cards—so new that the Moroccan state would not have gotten around to tapping them. I had likewise, and as instructed by Naseem, got myself a new local SIM just before ascending the coach (I had arrived in Marrakesh airport a day before and had been shadowed from the moment I crossed border control). Mohammed and Saleh could see me, but I could not see them. The few town spies that were up (my predawn arrival time was no coincidence) could see me. Abdelhay, Saleh, and Naseem had to get me lost. The former two did so by directing me—via Naseem and his London-based phone—this way and that way around the streets of Tarfaya until I lost my shadows.

It is possible to visit occupied Western Sahara as a tourist. In fact, Morocco promotes Dakhla, in its southern provinces, as a kite-surfing hotspot. Tourism is forcefully developed and has a royal connection. A new solar farm near the bay is conveniently built near hotels owned by Moroccan Princess Lalla.[6] What tourists may not realize is that they are watched. As long as they do not stray into a suburb populated by Saharawis or ask about Western Sahara in a public place, tourists probably will not notice the eyes following them. It is a different story if you arrive with an existing profile. Several checkpoints—ten perhaps—between Agadir and El Aaiún allow police and gendarmes ample opportunity to view the passports of tourists and check that they have no history of engagement with the Western Sahara question. If you travel at night, though, lethargic authority representatives might neglect their duties.

On my first attempted visit to Western Sahara, the sun did not rise until the last checkpoint before El Aaiún. No one checked my passport until then. Unfortunately for me and for one of my Norwegian colleagues (the other Norwegian colleague was of African heritage, sat far from us on the bus disguised as a Moroccan woman, and made it to El Aaiún to carry out her research on human rights abuses), after our passports were checked, we were removed from the bus and taken for several hours of questioning in the nearby station.

Eventually, plain-clothes police forced my colleague Ingrid[7] and me into a car. We were driven around for ten hours in a convoy of three plain-clothes police vehicles before being left in a car park on the outskirts of Agadir. From there, we were continuously intimidated as we slowly traveled north to catch a ferry to Spain to meet Saharawi communities there. By *intimidated*, I mean both rooms on either side of ours were very obviously booked and inhabited by plain-clothes police, one of whom would put a chair outside our room and follow us down corridors to the bathroom during the night. When outside of hostels, we were questioned every ten minutes as to why we had tried to travel to Western Sahara by people that seemed, to us, to be civilians. Were they all plain-clothes police? Or civilian informers? There was only one café where we were not questioned. There, a Moroccan woman sneaked us a note stating, "Be careful. There are bad men waiting for you outside."

We debated daily whether to attempt to meet with the various Saharawi students requesting a talk with us on the way to Spain. The Saharawi case is so absent in international media, human rights, and academic studies that Saharawi activists are keen to take advantage of interested visitors. Saharawis say that being seen with Global Northern activists also offers some protection—police might be more cautious about mistreating those who have campaigner friends from abroad. But still—hence our dilemma—there is surely always some risk for Saharawis and Moroccans willing to speak out against the Moroccan occupation. Meanwhile, our would-be host in El Aaiún called us each day as we traveled north and persuaded us to come back. We should have phoned him from the checkpoint, he told us. He is a good negotiator, and he could have persuaded the police to let us in.

What was the right thing to do? I pushed guilt into a dark corner, but it was still there, eyeing me all day, each day. In Rabat, we met the nephew of a Saharawi friend of mine. He took us on a tour of Mohammed V's mausoleum and to the best patisserie in town. We talked about his work on resource exploitation. Ingrid and I had met his cousin Hayat[8] in Agadir a few days before. Our picture together in the train station café afterward appeared on the front page of my hometown's local newspaper, along with a feature on the Western Sahara conflict. It also made it to the front page of a leading Norwegian national newspaper. "That's why we want to meet you, we need to get our message out," said Hayat. "We need people in powerful countries to know what is happening to us." Another day, in another city, we met Ahmed[9] and his friends, one of whom had his toe cut off during police torture in retribution for his human rights activism. The last meeting on our way out of Morocco was in Marrakesh: a police-interrupted breakfast with some Saharawi students. A policeman shouted at Ingrid and me, "Leave immediately, or we will beat you up!"

Later, in London, I discuss the experience with Naseem: "You should go back. We need international witnesses." The strategy of the Saharawis in the occupied territory relies on legitimacy and therefore on *people to observe that legitimacy*. Scholars of nonviolent resistance struggles identify sources of power and tactics that can serve as weapons for resistance movement activists.[10] The extent to which a movement depends on each of these weapons changes according to context. Although strikes, for example, have proven to be an effective tactic in various struggles, such a tool is of little use to Saharawis since they are underemployed and are heavily outnumbered by Moroccan settlers.[11] However, undermining the legitimacy of the Moroccan regime by exposing its violence is a key—if not the key—weapon of the Saharawi nonviolent anti-occupation movement. The legitimacy of the Saharawi struggle, *if only it were known*, could attract the support of an international civil society, which was so integral to the success of the East Timorese anti-occupation project and indeed to the end of apartheid in South Africa. It could also convert some corners of Moroccan civil society to support the Saharawi viewpoint. But securing the Saharawi struggle's legitimacy by exposing the aggressor's violence requires an international audience or at least a witness who can communicate the situation to a listening international audience.[12] Nothing is exposed if there is no audience to the exposition. A factor in explaining why Western Sahara is one of the most unknown settler-colonial struggles in the world is the largely successful media blockade that the Moroccan regime has maintained over Western Sahara. In the absence of the international media, then, other possible witnesses, such as researchers for human rights NGOs, solidarity activists, and academics become potential conduits for Saharawi nonviolent activists.

Marieke J. Hopman points out that the 2011 Moroccan Constitution effectively bans research on human rights in occupied Western Sahara.[13] The impact of this ban, argues Hopman, is that the international community remains completely blind to human rights violations in the region. Therefore, covert research on human rights abuses is not only ethically acceptable but also, in Hopman's words and with her emphasis, "*ethically required*."[14] The

Saharawi diaspora and their allies work hard to assist individuals with an interest in human rights, resistance movements, and/or the Western Sahara conflict in general to get into the occupied territory under the radar of the Moroccan authorities. Few international organizations make it into the occupied territory to report on the situation, and, if they do, they are forbidden, or at best discouraged, from returning. For example, Moroccan police detained the North Africa desk officer of Human Rights Watch when he was working on the 2010 Gdeim Izik uprising and beat him.[15] Parliamentary delegations from all over the world, as well as journalists, are routinely denied entry or quickly deported. In 2015, US *Democracy Now!* became the latest (at the time of writing) Global Northern media agency to access occupied Western Sahara but only managed four days in the territory. The Norwegian Support Committee sends dozens of delegations every year. Their usual forced deportation creates media interest in the Western Sahara issue. It is a way of raising awareness and support for campaigning in Norway, as well as showing Saharawis that there are witnesses to their suffering and that people elsewhere are in solidarity with their struggle. Similarly, the Spain-based Saharawi/Spanish-run group Sahara Thawra aims to send a constant stream of activists to the occupied zone so that there is always a Spanish witness to regime atrocities.[16] A similar US-based NGO organizes solidarity delegations to occupied Western Sahara embedded in scholar Gene Sharp's research on nonviolent action.[17] For sure, I spend time campaigning and use my research for such purposes. A main aim of my trips to the occupied territory was to document gendered human rights abuses for lobbying in partnership with Saharawi activists at UK, Norwegian, EU, and UN levels. But (with participants' permission) I also used some data in academic publications, including in the book that you hold in your hands or see on your screen. It is therefore—insofar as peer-reviewed publications are essential for an academic's progression—undeniable that my work on Western Sahara also furthers my career.

Back in Tarfaya, guilt bites again and is joined by his friend fear. I sprawl forward, and the door is opened for me by a smiling man pointing toward the space beneath the dashboard. I shove my backpack at him. It disappears to the backseats. I crawl in, folding myself up.

A portrait photograph swings on a string from the rearview mirror. It is El Wali, founder of the Polisario and national hero. Who else would it be? El Wali's resolute expression stares out to the desert in black and white. We exit Tarfaya town and leave the road for the open desert. Abdelhay and Saleh look considerably more relaxed than I feel. Saleh says, "Everything they [the Moroccan authorities] could do has been done to us already. It's impossible to feel fear anymore."[18] I do not know much about Saleh other than that he is a childhood pal of my good friend Naseem, but I know Abdelhay's family name well. Several of his relatives have been victims of forced disappearances and appear in reports by Amnesty International and other human rights NGOs.

Saleh turns up the volume on the cassette player. The song "Sahara ma timba" emerges. Its lyrics come from a Fatma Brahim poem composed as Morocco napalmed Um Dreiga refugee camp in 1975:

> The Sahara,
> my brothers and sisters,
> is not for sale.
> . . . .
> The phosphates you desire,
> they will cause you harm,
> I swear by God, you will not buy them,
> Not even if there is temptation to sell.[19]

> يا خوتي الصحرا، يا لوليلا
> وابدا ما تنباع، ما تنباع
> . . . .
> والفوسفاط تبغوه، يا لوليلا
> شايطكم بذراع، ما تنباع
> والله ما تشروه، يا لوليلا
> لو عاد لينباع، ما تنباع

We park at a lonely, small bungalow in the desert. It has a view of the Foum el Oud-Tarfaya road, the only tarmacked road that follows the coastline down to El Aaiún. Bashir[20] opens the door and ushers us into the living room, where the coals are already glowing on the tea brazier. The three men embrace each other warmly and ask after each other's relatives. Saleh interrupts the greetings to explain our stop: "We can wait here until the gendarmes get sick of looking for you in all the car boots." Sure enough, when I peer out the window, I can see a checkpoint in the distance where uniformed men are pulling over the few vehicles that pass.

Bashir lives alone—and is lonely and bored. "I've no future here. Jobs are only for settlers. I am trying to move to the Canary Islands," he tells me as he splashes sweet tea from one glass to another repeatedly, creating a foam that catches any sand particles. "I've tried before. There were a few of us in a fishing boat, but Moroccan police patrol the coast—they shot at us." Luckily, Bashir managed to swim back to shore. Morocco is the southern border guard of Fortress Europe. It sees fit to shoot to kill when it comes to would-be emigrants.[21]

We pass the time with a tasting session. Abdelhay has brought milk from his nomadic family's camels. Each animal's milk is in its own separate small bag. He shows me (cutting the corner of each bag so that I can drink) that the flavor of each animal's milk is unique—some sweet, some with a yogurt-like tang. Saleh peers out the window now and then. Finally, we go outside. "They've stopped checking cars now. Let's take some photos together for Naseem. Put these on." He hands me a Saharawi man's camouflage suit.

I hoped we might avoid complications by going straight from Bashir's place to my family's place in El Aaiún as the crow flies—no stops.[22] But Abdelhay and Saleh insist on one detour. Not far from the Western Saharan capital, they call me up from beneath the dashboard to gaze at something they want documented by anyone from abroad: a Siemens wind farm, in a very blustery Tarfaya.

The wind also blew hard on a March morning in 2013, right here, right by where the car is passing now. Sixty-eight-year-old Saharawi Moulmoumnin Salek Abdesamad bent into the wind as she walked to her pasture. When she arrived, a large fence blocked access. As her hungry goats bleated at her heels, she peeked through the mesh to see that her land had been perforated by large holes. The Moroccan partner of French GDF Suez and German Siemens had appropriated the grazing area, with no prior warning, for the construction of the 300-MW Tarfaya wind factory.[23] When Moulmoumnin complained, reports NGO Western Sahara Resource Watch, she was arrested and taken to the police station. "There is no sense in opposing the king's project," she was told.[24] Her case was not unique. Saharawis tell me of three other Saharawi families who also lost land to this wind farm.[25] Two years later, during the question-and-answer session after a presentation of mine at a conference at Durham University, a Siemens employee in attendance raised his hand to deny all of this and negate the existence of any Siemens turbines in occupied Western Sahara itself.

In 2014, a little south of here, in occupied Western Sahara, Moroccan police interrupted Elghalia Boujamaa's breakfast. For several months beforehand, her family had been under pressure to abandon their home because it sat on the would-be path of French company Alstom's electricity line. On October 22, 2014, Elghalia and her family members were finally permanently removed from the house, after a severe beating.[26] Energy infrastructure in Moroccan-occupied Western Sahara is tangled up with oppressive political violence. Recent scholarship on energy and power highlights how energy mediates politics and power relations globally.[27] Although in some contexts social movements push for emancipatory energy transitions,[28] there is a large body of research on less-than-progressive uses of energy infrastructure. Using the case of Mozambique, Kirshner, Castán Broto, and Baptista highlight how perceptions of energy landscapes reflect social relations of domination.[29]

Other studies focusing on Global Southern contexts highlight the colonial implications of such power politics. For example, Charis Enns and Brock Bersaglio argue that large infrastructure projects in East Africa are reproducing colonial legacies.[30] Similarly, Power et al. argue that socially and spatially uneven energy infrastructural developments in Mozambique and South Africa are at risk of perpetuating a colonial electrical geography, given that they favor corporate and elite clients over households and communities.[31] Energy developments are also contributing to settler-colonial conflicts in settings as diverse as Mexico and Palestine.[32] My concern in this chapter is with the settler coloniality of the Moroccan energoregime in occupied Western Sahara.

David Lloyd and Patrick Wolfe powerfully argue that settler colonialism is not an early (and past) stage of capitalist expansion but rather a current conduit to, and tool of, the neoliberal world order of capital accumulation.[33] I argue in this chapter that the reverse is also true: the neoliberal, capitalist world order is likewise a servant to settler colonialism. I move away from exploring the *difference* of settler colonialism and look at how other colonial actors—specifically a multinational corporation—support, reinforce, and further the work of the settler-colonial mission. First, I focus on how wind and solar developments in occupied Western Sahara are mobilized internationally in support of Morocco's occupation. Second, using the case of the main (at the time of writing) European actor in occupied Western Sahara's wind factories, Siemens, I unravel the wind-infused narratives and imaginaries that the corporation uses to justify, to external audiences, its presence in occupied Western Sahara. Doing so reveals the colonial mindset—remarkably similar to that of the European and American shipwreck survivors and Spanish colonists—on which Siemens plays, and which it furthers, as well as its participation in Morocco's settler-colonial strategies.

Energy Diplomacy, Energy Colonialism, and Foreign Corporate Allies

If you board a coach from one of the big Moroccan cities southbound to Western Sahara, you will pass through a town called *Guelmim*. Two hundred years ago or so, several European and North American captive seamen passed through here. Then, it was a settlement of clay houses and souks, a crossroads for Trans-Saharan caravan traffic and a convenient stopping place for sailor-captors like Charles Cochelet on their route to redemption in Mogador. Today,

Guelmim is not within the legal borders of the non-self-governing territory of Western Sahara, but it was once part of ancient Saharawi lands. It still today has a large Saharawi population. If you traveled through Guelmim in late April or early May 2011, you might have seen three makeshift tents through your window. At the roadside, there would have been rugs and awnings hanging from poles, rocks out front to prevent would-be hit-and-runs, and a huge white banner reading, in Arabic, in the red, green, and black ink of the Saharawi flag, "A peaceful sit-in by Saharawi nationals of Guelmim to demand their rights to work opportunities and to live in freedom and dignity."[34] If this sight provoked your curiosity and you dismounted the coach with the intention of approaching the tents, you probably would have been prevented from doing so. Plainclothes police would most likely have put you in the back of a car and driven you north, as far away as possible from Saharawis and Western Sahara, for your own good. But let's suppose for a minute that you somehow made it inside the tent. You would have found six young men on hunger strike with several friends tending to them. They would have looked gaunt, with sunken cheeks and wide eyes. They would be lying down, clutching pieces of paper with their demands translated into English: "PCS is plundering our land" (PCS refers to the US-Canadian fertilizer producer PotashCorp), "EU, respect our rights!" "San Leon, leave!" "Please cease the stealing!"[35]

At times, Saharawis have held monthly demonstrations like this one in Guelmim against natural resource exploitation.[36] The activists draw our attention to phosphates, which are mined thanks to an adjacent Siemens wind farm, to the EU, which imports agricultural and fish products from occupied Western Sahara, and to San Leon, the Irish-British energy company. In 2015, San Leon became the first company to drill for oil onshore Western Sahara. Oil companies have not (yet) found oil in commercial quantities in Western Sahara. Since fossil fuels elude Morocco, the kingdom imports most of its energy (90 percent, according to 2019 data).[37] It makes itself—and I wish to draw attention to the reflexive verb here—dependent on occupied Western Sahara for its future renewable energy independence. This reflects similar situations in other colonial contexts—for example, Gaza's ties to Israel via electrical and water infrastructure.[38] However, while Israel forces Gaza's dependence on itself through infrastructural violence, Morocco strategically chooses to make itself partially dependent on Western Sahara for energy.

Western Sahara is connected to the Moroccan grid via an interconnection in the capital El Aaiún. A 400-kV power interconnection between El Aaiún and, to the south, Dakhla is in progress.[39] Morocco hopes that its grid will eventually connect to the Mauritanian one via Dakhla city,[40] through a power line with a 400-kV transmission capacity.[41] Morocco plans to export renewable energy to West Africa via this link, as well as to the European Internal Market (a roadmap to this end was signed at the 2022 United Nations Climate Change Conference of the Parties [COP22]) via existing and planned submarine connections with Spain, Portugal, and, as announced in 2020, possibly the UK via a 3GW submarine cable between the UK (thanks to UK company X-Links, backed by UK Octopus Energy) and the Sahara, which would generate energy to meet 6 percent of the UK's demand.[42] The implications for Saharawis' right to self-determination are huge. Aside from the financial benefits for Morocco, these planned energy exports would make not only Morocco but also the European and West African energy markets partially dependent on energy generated in occupied Western Sahara. This manufactured dependence is politically charged. Energy developments in occupied Western Sahara stake a colonial claim on illegally occupied land and implicate foreign powers in this claim. The Moroccan regime uses energy to recruit allies for its colonial project.

With regard to the Mauritanian connection, this would provide access to the African market, which is "very promising [for Morocco] since it is relatively poorly electrified and represents real investment opportunities for future years."[43] Wind and solar energy is at the heart of Morocco's wider foreign policy in Africa. While Morocco left the Organisation of African Unity in anger in 1984 when the Saharawi Arab Democratic Republic was admitted as a member state, it joined its successor organization, the African Union, in 2017. The kingdom now seeks to augment its power in the African arena and is using the promise of energy as one form of soft power.[44] Morocco has modeled itself as "the African leader in renewable energy in Africa."[45] In 2021, it announced a "Coalition for Sustainable Energy Access" in partnership with Ethiopia as well as a phosphates export deal with the latter in exchange for Ethiopian support of Morocco's war against Polisario.[46] Simultaneously, Morocco is planning a Nigeria–Morocco gas pipeline, which would connect all countries of the West African coast with Nigerian gas. This has softened Nigeria's pro-Polisario stance.[47] ONEE, the Moroccan state's electricity and water company, has already provided technical assistance and expertise for electrical infrastructure projects in Senegal, Gambia, Sierra Leone, Guinea Conakry, Cape Verde, Chad, and Mali.[48] Morocco's publication launched at COP22, *ONEE au Maroc et en Afrique*, underlines the place of energy in realizing the kingdom's diplomatic efforts in securing support for its occupation in traditionally pro-Saharawi independence, pro-Polisario sub-Saharan Africa.[49] Although the most widely used definition of "energy diplomacy" is the use, by exporters, of foreign policy to secure energy markets abroad,[50] what we see in Morocco's case is the promise of future contributions to neighboring regions' energy supply in return for recognition of its illegal occupation.

Wind and solar factories also do diplomatic work in support of Morocco's illegal occupation well beyond those countries who will benefit from energy exports. Morocco is routinely celebrated as a pioneer in the energy transition, but few of the articles and documentaries that laud the kingdom talk of the huge human cost of this exemplary transition or highlight that much of the energy does not come from Morocco at all. Alicia Fernandez Camporro has illustrated how Moroccan King Mohammed VI uses terra nullius and greenwashing narratives to justify his appropriation of Saharawi lands for large-scale energy developments.[51] Her discourse analysis of Mohammed VI's recent speeches on world stages demonstrates how Mohammed VI operationalizes ideas of terra nullius to paint his southern provinces as underdeveloped and in need of growth—wants that can be met through environmentally friendly energy developments. Besides inflating his own bank account and creating further infrastructure to bolster his kingdom's occupation, Mohammed VI gains international prestige for his leadership in green energy and for increased opportunities to access green finance. And, as Fernandez Camporro crucially argues, the king's discourse on wind and solar developments strengthen wider discourse of *national unity*, designed to foster Moroccan public support for the continued occupation of Western Sahara and to paint the occupation as irreversible in the eyes of foreign onlookers. Siemens follows Mohammed VI in operationalizing discourses reminiscent of Europe's civilizing mission, which allegedly involved the altruistic modernization of empty or underused lands left to waste and the gift of education and technology to backward Indigenous peoples. In international-relations terms, Morocco's investments in wind and solar factories give the kingdom a new source of soft power allowing it to move closer to becoming a global player.[52] Whether energy in occupied Western Sahara is produced from renewable sources or fossil fuels makes little difference to the overall coloniality and oppressive nature of the energy system. However, the presence of wind and solar factories has the advantage of casting a favorable light on Morocco due to the power of greenwashing discourses.[53]

Based on Morocco's published plans, Western Sahara Resource Watch estimates that by 2030, 47.2 percent of Morocco's total wind capacity and 32.64 percent of its solar capacity will be sourced in occupied Western Sahara.[54] Almost the entire area of Moroccan-occupied Western Sahara is to be surveyed (by Portuguese company Gesto) in terms of its potential for generating geothermal energy.[55] Of five wind farms in occupied Western Sahara, four belong to the portfolio of the Moroccan king's energy company NAREVA, including the 50W Foum el Oued farm. Much like the ironic case of wind farms powering the highly contested (by the local population) Kudankulam Nuclear Power Plant in India, the Foum el Oued farm currently provides 100 percent of the energy needed to power the neighboring Fosbucraa mine,[56] a key source of phosphates globally.[57] The existing Aftissat wind farm is also used for industry. What we see, then, is a favoring of corporate clients in terms of energy provided by existing wind farms; this is also the case with large energy developments in Mozambique and South Africa.[58] Of solar infrastructure, the newest planned addition at the time of writing will be in Dakhla, a popular destination for kite surfers. The farm will be built on the opposite side of the bay to the city's hotels, which are owned by the king's cousin Princess Lalla, so as not to spoil the tourists' view, and will conveniently be adjacent to industrial greenhouses owned by the king, members of the *makhzen*, and French conglomerates.[59]

Research on the neocolonial and pro-elite implications of the energy system in Morocco itself are relevant here. Jawal Moustakbal shows how a 2015 Moroccan law took up the concept of availability-based payments. Through this model, public sector body ONEE must buy the entire output that private concessionaires produce, regardless of actual needs, at substantial cost to Moroccan and Saharawi taxpayers and at substantial profit for foreign corporations (Moustakbal names Siemens and Siemens Gamesa) and royal shareholders who own many of the energy developments, including all but one of the existing wind factories in Western Sahara.[60]

There are hundreds of foreign businesses involved in the exploitation of occupied Western Sahara's natural resources, but some are more deeply implicated than others. Saharawis' corporate enemy number one is arguably Siemens since the company is involved in most of the wind farm developments in the country. Siemens has been described as "the most important European player currently operating in Western Sahara on a Moroccan contract."[61] Energy exploitation constitutes a unique threat to Saharawi hopes for independence for two key reasons: it is currently the only form of resource exploitation that physically connects the two territories of Western Sahara and Morocco (via powerlines, cables, and so on), and it gives Morocco the long-coveted prize of energy self-sufficiency. Siemens is like other companies exploiting the territory's resources in two main ways: it offers recognition of Morocco's illegal occupation of Western Sahara, and it mirrors and fosters Morocco's own colonial discourses on Western Sahara.

All colonialism entails forms of exogenous domination and unequal power relations. The term is commonly used today as a broad concept to describe

European political domination and control over much of the rest of the world from the sixteenth century to the late twentieth century.[62] But colonialism can nevertheless take different forms. In this chapter, I discuss *"classic" colonialism* (I borrow *classic* from Erick W. Steinman) versus *settler colonialism*, and both versus *neocolonialism*.[63]

Researchers working in settler-colonial studies have shown the importance of considering the specificities of *settler* colonialism, as opposed to other colonial practices.[64] Settler colonialists, "outsiders who come to stay and establish territorialised sovereign political orders,"[65] seek to eliminate the Indigenous population,[66] whereas nonsettler colonialists seek to maintain the Indigenous population to exploit it. Lorenzo Veracini explains the difference as follows: a colonist meets an Indigenous person and says, "You work for me." A *settler* colonist meets an Indigenous person and says, "Go away."[67] *Go away* can imply various strategies of elimination: physical elimination or displacement, erasing Indigenous cultural practices, or absorbing, assimilating, or amalgamating the Indigenous population into the settler one.[68] In other words, classic colonial regimes attempt to perpetuate the hierarchical colonizer-colonized divide, while settler colonialism seeks to eliminate all colonial determinants. The ultimate fulfilment of settler colonialism is to extinguish itself by a discontinuation, or total assimilation, of Indigenous claims and subjectivities.[69] Veracini says that settler colonists "repress, co-opt, and extinguish indigenous alterities, and productively manage ethnic diversity. By the end of this trajectory, they claim to be no longer settler colonial."[70]

Spanish colonialism in Western Sahara, or Spanish Sahara, as it was known, could be considered an example of classic colonialism. Although thousands of Spaniards moved to "their" Sahara, discontinuation of the colonial regime was always a conceptual possibility, and indeed, when the Spanish state left Western Sahara in 1975, so did the Spanish population. Spain attempted to profit from its colony economically. The aim of Spanish colonialism there was not to *replace* the Indigenous population but to exploit the territory's resources and to exploit the Indigenous Saharawis to this end, for example, through cheap labor in the phosphates and fisheries sectors. To go back to Veracini's example, Spain did not say, "Go away" to the Saharawis but rather, "Work for me."

Morocco's approach in Western Sahara is typical of settler colonialism. Morocco has attempted to eliminate the Saharawi population *physically* via displacement (through war), discrimination (so as to force Saharawis to emigrate), gendered policies of violence that prevent the most nationalist Saharawis from reproducing (through forced sterilization, for example, or preventing access to jobs that could enable someone financially to sustain a family),[71] and genocide (through napalming and slaughtering fleeing refugees), and *culturally* by way of oppression of all Saharawi identity markers. In contravention of Article 49 of the Fourth Geneva Convention, and in a bid to integrate Western Sahara into Morocco proper, Morocco used the 1991–2020 stalemate to move an ever-increasing settler population into occupied Western Sahara. Moroccan settlers account for an estimated 75 to 80 percent of the territory's population.[72] Despite this, most experts agree that the overwhelming majority of Indigenous Saharawis support the independence cause.[73]

But Morocco also facilitates *neocolonialism* in occupied Western Sahara. Neocolonialism emerged in the post-World War II period, following the waves of political independence movements in the Global South in the fifties, sixties, and seventies. As Europe's formal colonial rule in several Asian and African countries came to an end, neocolonialism stepped into action. *Neocolonialism* was the name given to the process by which Europe, and settler colonists of European origin in North America and Australasia, continued to exploit, make economic profit from, and advance capitalism in the ex-colonies. This process has involved activities such as, for example, questionable aid initiatives, unfair trade practices and finance mechanisms, export of natural resources, fanning the flames of political and military conflicts in the Global South, establishing military bases and co-opting international organizations, and installing mechanisms to further neocolonial and neoliberal policies.[74] In occupied Western Sahara, the exploitation of natural resources by external corporations, states, and groups of states (e.g., the European Union) in partnership with Morocco are a form of neocolonialism.

Given the power of greenwashing for the Moroccan occupation, and its role in Siemens's self-promotion, it is important to highlight the "environmental violence" of the Moroccan occupation.[75] As Siemens supports the occupation, it is complicit in the occupation's environmental crimes. The SADR government highlights in its indicative Nationally Determined Contribution (iNDC) to the Paris Climate Agreements that "the most dramatic and direct impact of the conflict is the interruption of drainage systems by the Berm."[76] One can observe this damage on satellite imagery from several points along the berm, or military wall. The wall cuts through wadis, blocking the passage of water into the liberated territory and thereby destroying plant and animal life. As the SADR government explains, "These locations thus experience the double impact of physical diversion of surface runoff coupled with increased evapotranspiration resulting from higher temperatures. In this way, the physical manifestations of the conflict combine with climate change to amplify risks to ecology and biodiversity. The Berm and the minefields that run along its length also pose a risk to wild fauna."[77]

The occupation's environmental crimes are linked to state efforts to prevent what Saharawis consider to be a defining feature of their identity: nomadism. The severity of these crimes becomes clearer if read through Nixon's concept of slow violence, which he defines as "a violence that occurs gradually or out of sight, a violence of delayed destruction that is dispersed across time and space, an attritional violence that is typically not viewed as violence at all."[78] Also relevant is Crook and Short's theory of genocide-ecocide. They explain that "by ecologically induced genocide we mean any ecologically destructive practice, or process, that forcibly controls indigenous and place-based peoples' interaction with, ejects them from or prejudices or precludes the enjoyment of, their land and the local ecosystems."[79] To remove the ability of a place-based community such as the Saharawi from enjoying or living with their land, through environmental degradation or other means, is a form of slow violence and, in the case of Moroccan policy, a form of genocide-ecocide. For Short, "neoliberal capitalist landgrabs . . . through the annexation of indigenous land" are the principle vectors of ecologically induced genocide.[80] In the case of Western Sahara, one might read Morocco's occupation as a land grab from Saharawis as a whole, but we should also read individual land grabs, such as the aforementioned stealing of Moulmoumnin Salek Abdesamad's grazing land in order to make way for a Siemens farm, through the same lens.

Morocco's settler-colonialist approach seeks to make impossible the Saharawi nomadic way of life. Oppressing nomadism during the first war (1975–1991) involved the mass slaughter of nomads' camels to deprive them (and Polisario) of a key resource,[81] as well as blocking nomads' wells.[82] At the outset of the second war in 2020, one of Morocco's first actions was another slaughter of Saharawi nomads' camels. The millions of land mines scattered across the territory threaten nomads and their animals, causing several deaths every year.[83] As the UN's Special Rapporteur for cultural rights has noted, these mines jeopardize the traditional life of nomads, and Saharawi victims of mine accidents face difficulties in accessing police reports to claim compensation. She also noted that international NGOs are not allowed to assist in demining efforts.[84] Since 2010, Morocco has prohibited the pitching of Saharawi *āḫīām* in the desert. Around the same time (2010), the Moroccan army implemented a ban on entering the roughly 50-kilometer-square Um Draiga region, an area of particularly rich and fertile grazing lands.[85] Natural resource exploitation, including via large industrial greenhouses as well as wind and solar farms, further shrinks grazing lands. Since mobility is crucial to the health of the camels, the reduced grazing lands threaten the survival of nomadic pastoralism.[86]

The military occupation and the natural resource exploitation that accompanies it also threaten the Saharawis' traditional existence via the destruction of the precious ecosystems that make nomadism possible. The Saharawi Association for the Monitoring of the Resources and the Protection of the Environment (AMRPENWS) claims that the few green areas of occupied Western Sahara are being destroyed by Moroccan military personnel, some of whom chop down trees for firewood, for producing charcoal, or for wood to sell to Turkish baths and bakeries. AMRPENWS also decries the excessive use of pesticides in farms near Dakhla and Foum el Oued, the lack of recycling and waste management infrastructure, the pollution of rivers and underground water sources due to pollution from trade effluents, and the poaching of endangered species.[87] Western Sahara Resource Watch has reported on the depletion of underground desert wells; huge industrial greenhouses (likely to be powered by solar farms in the future, as mentioned above) owned by the Moroccan king and French conglomerates pump underground water to irrigate their produce.[88] Also pertinent is the potential desertification and biodiversity loss caused by Siemens-wind-farm-powered phosphates mines.[89] Even if one was to apply what Lohman terms "white climate" logics to the case of wind and solar factories in occupied Western Sahara and argue that the environmental benefits excuse the developments' implication in the misery of the Saharawi people, one might have quite a task in proving that said benefits exist at all.

Aiding and Abetting Colonialism

"In Africa, for Africa. In Morocco, for Morocco"—so go the campaign slogans of the Siemens Africa office and the Siemens Morocco office, respectively. They suggest that the German multinational is in the continent for selfless reasons, to help the backward Africans. With such mottos, Siemens adopts, to borrow Apollo Amoko's words, a "missionary position" apropos to its presence in Africa.[90] That is, Siemens is a metonymic dispersal or *white presence* serving as a stand-in for the missionary or *white savior* of classic colonialism.[91] Befitting of such a role, when Siemens speaks of its work in the continent, and more specifically in Morocco (and occupied Western Sahara), it does so with the tongue of a nineteenth-century European colonizer. In this section, I analyze Siemens's narration of—or its discursive attempts to legitimize—its own presence and activities in the region. My methodology here has its edifice in critical discourse analysis (CDA), and more specifically in Laclau and

Mouffe's political take on discourse analysis, which sees all social practices, linguistic and nonlinguistic, including written material, organizations, and historical events, as texts for analysis.[92] CDA is about the power of language. It is a methodological tool forged to reveal how discourse creates and maintains discrimination and unequal power relations while simultaneously obfuscating and naturalizing them.[93]

My data is Siemens-authored texts: public-facing publications, communications, and presentations, and also built energy infrastructure itself. Settler societies are "premised on the elimination of native societies."[94] Reading infrastructure as a text, I argue that Siemens facilitates this elimination through land grabs and erection of mills—quite literally powering colonialism. I also look at Siemens's discursive *nullifying* of Western Sahara and of Saharawis, which draws on and continues Morocco's work of cultural appropriation. In terms of the scope of the materials that I use, I go back to 2009 and the DESERTEC project (outlined below), when Siemens first envisaged working in Western Sahara. I use external-facing, publicly available publications, presentations, and communications on Morocco since Siemens avoids referring to Western Sahara as such, preferring to call it *Morocco* unless asked a specific question on its activities there. All the material I use is published in English since this is the official and main working language of Siemens. Below, I show how Siemens mimics colonial terra nullius narratives; reproduces the civilizing mission, or civilization versus barbarism, discourses of the European colonial era; greenwashes its participation in the ecological disaster that is the occupation; and participates in Morocco's wider settler-colonialist appropriation of Saharawi culture.

Siemens is a German multinational conglomerate working in the fields of health care, infrastructure, industry, and energy. Since 2017, its operations in occupied Western Sahara have been led by Spain-based Siemens Gamesa, which manufactures wind turbines and is a provider of "wind power solutions."[95] Siemens Gamesa is largely owned by Siemens, and German Siemens was responsible for operations in occupied Western Sahara until 2017. Therefore, when I refer to Siemens throughout the remainder of chapter, I do not differentiate between the original German company and its Spanish offspring.

At the time of writing, the only searchable reference to Western Sahara on Siemens's global website is a shareholder counterproposal submission from the Cologne-based NGO Dachverband der Kritischen Aktionärinnen und Aktionäre, which denounces Siemens's questionable practices around the world, from its investment in the occupation of Western Sahara to alleged complicity in human rights violations in Honduras and Mexico.[96] Indeed, Siemens is publicly silent on Western Sahara. While their wind developments in Morocco (and elsewhere in the world) are promoted with several pages, photographs, interviews with local staff and beneficiaries, videos, and tweets and glossy reports, developments in Western Sahara receive considerably less coverage. At the time of writing, I was not able to find a single press release or other type of public-facing communication concerning its involvement in wind developments at Tiskrad, near El Aaiún, or Aftissat, near Boujdour. Previously, I had found only one communication, a press release, focusing on its earliest project in occupied Western Sahara, which Siemens described as being located in Morocco.[97] However, this press release has disappeared from the Siemens website by the time of writing.[98] Siemens answers questions from NGOs and Saharawis on its involvement in occupied Western Sahara only occasionally—and even then, evasively. Siemens's public discourses on its investment in occupied Western Sahara can be summarized in two words: "No comment." For Siemens, discursively, there is no Western Sahara. The corporation renders the country nullius.

Greenwashing is the use of marketing to portray an organization's products, activities, or policies as environmentally friendly when they are not. The Greenwashing Index, an online mechanism for flagging incidences of greenwashing, explains the concept using the "classic example [of] an energy company that runs an advertising campaign touting a 'green' technology they're working on—but that 'green' technology represents only a sliver of the company's otherwise not-so-green business."[99] The aforementioned single press release on the Foum El Oued farm gives Siemens the chance to highlight its contribution to the "emerging African wind market" and to "one of the most promising sectors for renewable energy generation in Morocco," its role in creating jobs for Africans in said market, and its part in empowering Africans to reduce carbon dioxide emissions through Siemens's "ecofriendly technologies" and its "environment portfolio."[100] Its endeavors in occupied Western Sahara are presented as supplying socioeconomic opportunities while respecting the environment. If we focus on Siemens's articulation of its role in the wider *Moroccan Integrated Wind Energy Programme*, of which the various wind farms in occupied Western Sahara form a part, there is remarkable consistency in the environmental language used. However, if one focuses on the specific environmental impact of the Foum el Oued farm, and, from a world ecologies perspective, on the wider system of which the Western Sahara developments form a part, this narrative can be classed as greenwashing. Siemens makes almost three times as much revenue from its operations in the gas and power sector than it does in the wind and solar energy sector (throughout 2016 to 2017,

7.1 percent of its revenues came from Siemens Gamesa Renewable Energy, whereas 20.1 percent came from power and gas).[101] The fruits of the Foum el Oued wind farm directly undermine Siemens's environmental rhetoric. As mentioned above, the owners of the (extractive, nonrenewable) phosphate mine, the first source of phosphates worldwide, have claimed that 100 percent of the energy needed to run the mine comes from the Foum el Oued farm.[102] If we approach the Western Sahara case in world-ecology shoes, which collapse the nature/society dichotomy and see world ecology as the present and past of capitalism, then green industry exploitations of Western Sahara's resources are no different from those carried out in the name of the phosphates, agriculture, and other sectors.[103] They are tangled together in an ecology that causes droughts, dries up the desert, and destructs the ocean environment.[104]

Morocco has made major fishing ports of El Aaiún and Dakhla, with large processing factories in both cities. While industrial vessels deplete resources and use risky methods such as dredging, local Saharawi fishermen using traditional fishing methods are often denied licenses due to their political opinions, or they are left with ever-dwindling catches.[105] An independent evaluation of the fisheries sector in occupied Western Sahara, carried out on behalf of the European Commission, finds that European and Moroccan vessels have targeted endangered species such as sharks, rays, and skates; that almost all commercially viable fish stocks are either fully exploited, overexploited, or unclear in terms of their status; and that continued fishing there by EU vessels would further harm the fish stocks.[106] The Food and Agricultural Organization (FAO) has similarly reported that regional sardinella stocks are on the point of collapse.[107] Single named vessels have reportedly each discarded over one thousand tons of fish during a one-year period.[108] In 2017, a Moroccan captain burned himself to death in protest against what he saw as the corruption, injustices, and threats that riddled the fishing industry in occupied Western Sahara.[109] The situation of military occupation quickens and intensifies these symptoms.

Siemens presents Moroccans and Saharawis as one and the same. With remarkable consistency in vocabulary, in its promotional materials on work in Morocco (invariably including occupied Western Sahara), it talks of the jobs, energy, and other benefits provided for "the local population." Similarly, when asked by NGOs and Saharawis how it has ensured that its investments are in line with the wishes and interests of the Indigenous Saharawi people (a legal requirement for exploiting the natural resources of an occupied country), it refers to its consultations with "the local population." It ignores the issue that, for Saharawis, and for international law, the Moroccan settler population is different from the Indigenous Saharawi population. In the fashion of settler colonialism, Siemens ignores the existence of the latter as an entity separate from Morocco.

Siemens also assists Morocco in its project of settler colonialism by increasing infrastructure and therefore working toward the denial of future alternatives to the total elimination of Saharawi nationhood. Jabary Salamanca has highlighted how academic literature on international development has consistently, but misguidedly, celebrated the importance of infrastructure networks in the Global South, regarding them as *apolitical* indicators of progress and modernity that bring innumerable benefits to local communities.[110] Such literature overlooks, argues Jabary Salamanca, the unequal power relations that infrastructure can articulate and enact, and the ability of infrastructure to carry out territorialization on behalf of settler colonialists.[111] Infrastructural development is a discursive and acutely political practice as well as a material one.[112] "Infrastructures, cities and nation states are produced and transformed *together*,"[113] and infrastructure reproduces certain ideas about territorial sovereignty and the state.[114] Infrastructure *networks*, such as energy grids, play a particularly crucial role in the discursive construction of a territory (or nation) because they create connections between places in physical, economic, and political ways.[115] Just as in classic colonialism, governing infrastructure was a powerful means of controlling and disciplining the "natives" whose labor needed to be exploited.[116] As Cole Harris points out, the ability of settler colonialists to *dispossess* an Indigenous population similarly rests on the former's physical power coupled with its ability to create a supporting infrastructure.[117] Cole argues that colonizing infrastructure is given moral validation by drawing on colonial discourses of terra nullius and barbarism versus civilization—civilization being the Western-centric, capitalist ideal of economic growth and "modernity."[118] In other words, material infrastructure physically (literally) supports settler colonialism while simultaneously seeking its own moral justification on an ideological level through the propagation of colonial narratives. In the case of Moroccan-occupied Western Sahara, Siemens literally (physically) powers the occupation and literally (physically) connects occupied Western Sahara to Morocco proper via power lines. It also creates the conditions of possibility that allow the Moroccan state to cut power to the Saharawi-dominated suburbs at particular political moments, as is further explored in chapter 4.

Siemens dabbles in Orientalism and cultural appropriation, supporting the wider strategy of Morocco in the latter. Orientalism, as conceptualized by Edward Said, is the pervasive Western tradition of fetishizing and exoticizing

Arabs to create a binary between "civilized" West and "backward" East.[119] One might interpret the romanticizing of a wind factory with images of nomads' wandering camels as an example of Orientalist discourse. The background photograph to Siemens's pages on its Tarfaya wind developments, which were removed when operations in Western Sahara were taken over by Siemens Gamesa, depicted four camels in the foreground, standing in a line among desert rocks and scrub.[120] Behind them were a number of pristine white windmills. The colors were muted, allowing the mills, camels, desert, and clear sky to blend together into a beige and dimmed-blue soup. The positioning of the camels mirrored that of the mills. Aesthetically, the photograph was engineered to make the mills look at home in the desert landscape. The camels, who appear to have happened upon this vast wind farm during their desert travels, were suggestive of a traditionally *Saharawi* way of life: nomadic pastoralism. Siemens appropriated the camel as a trope for desert nomadism just as the Moroccan kingdom does more widely within its settler-colonialist strategy. The Saharawi lifestyle becomes associated with Morocco, and the image ironically suggested that the Siemens development and this traditional way of life can happily coexist.

It is important to highlight the significance of the use of camels in this picture. While attempting to make nomadism impossible for Saharawis, Morocco simultaneously appropriates nomadic culture. Volpato and Howard have chronicled, with useful detail, the extent to which the heritage of camel pastoralism is embedded within the Saharawi nationalist imaginary and used by Polisario as well as Saharawis living in occupied Western Sahara as a border marker with Moroccans. The latter, accordingly to Volpato and Howard, are relatively ignorant of camel pastoralism, husbandry, and health.[121] Saharawis use camels and camel milk (which is reportedly served to ex-political prisoners when they emerge from Moroccan-run jails) as symbols of ethnic and national identity.[122] As Volpato and Howard show, especially since 2008 or 2009, Morocco has attempted to reposition Saharawi nomadic heritage as its own through state aid for camel husbandry, free medicines and vaccines, funding for epidemiological studies, large events held in "Moroccan Sahara" celebrating the heritage of nomadic pastoralism (this goes hand in hand with an increasing use of the Saharawi *ḫaīma* as a symbol of Moroccan heritage), and the promotion among Moroccan settlers of the consumption of camel meat and milk, as well as the export of the latter abroad.[123] Similarly, Siemens occasionally accompanies promotional articles with photographs of implied Moroccan employees wearing traditional Saharawi clothing.[124] Siemens participates in settler colonialism by presenting Saharawi culture as Moroccan.

Creating a terra nullius is a precondition of capitalist development. As Franquesa explores in his ethnography of southern Catalans resisting wind developments, (renewable) capital justifies the need to intervene by constructing places and people as waste. For Franquesa's southern Catalans, to accept the construction of wind farms on their ancient farming lands is to accept that their beloved territory is a wasteland, "a place where the things others do not want are installed, where the resource is extracted, while profits and electricity go elsewhere."[125] As discussed in the preceding chapters, wind imaginaries of the shipwrecked and the Spanish colonialists informed the construction of Western Sahara as a terra nullius. Nowadays, the wind energy industry continues this work, evoking the wild desert winds in modern-day colonial constructions of Western Sahara as a terra nullius. Given the company's near silence on its work in Western Sahara itself, my analysis explores Siemens's discourse on its wider contributions to the *Moroccan Integrated Wind Energy Programme*, of which the developments in occupied Western Sahara form a part. The title of the Siemens in Morocco home page (until Siemens Gamesa took over) was "Turning the Wind's Energy into a Source of Economic Development. That's Ingenuity [*sic*] for Life." The page featured an article describing the Tarfaya wind farm ("Africa's largest onshore wind farm") and its contribution to carbon offsets. It also highlighted the high-tech prowess of Siemens: "Wind turbines have been specifically adapted to withstand the corrosive conditions of both the salty ocean winds, desert sandstorms and the year-round hot weather."[126] Here, Siemens built on preexisting discourses, hegemonic in Western public imagination (in part thanks to the discursive work of the mariners' accounts discussed in chapter 1), of the desert as scorching, inhospitable, and hostile in order to underline its technological acumen in overcoming this "challenging, very dry and dusty" context. As Hunold and Leitner point out, the desert as a lifeless, barren, and useless wilderness is a culturally dominant one in Western society.[127] Their discourse analysis of stakeholders' narrative strategies for gaining public endorsement of the Solar Grand Plan, a renewables venture in the Mojave Desert, provides useful lessons for the Western Sahara case. Hunold and Leitner show how investors in the Solar Grand Plan left intact, and built on, discourses already hegemonic in the public imaginary: deserts as of little use to anyone, as hostile, as empty, as apocalyptic, and as in need of "taming" by modern technology.[128] Likewise, John Beck's study of the North American deserts similarly identifies common tropes of emptiness and vacancy, thereby allowing desert spaces to become venues for (military and scientific) experimentation, construction projects, and economic development.[129]

The article on the Tarfaya wind farm ended with the line "Capturing the power of desert winds—and turning it into clean energy millions rely on. That's Ingenuity for life."[130] The idea of the wind as something that humans can harness, that nature is something to be dominated, is repeated in the corporate video that was embedded in the web page and that was still viewable online at the time of writing. Like the article and wider web page, the video is designed to promote Siemens via its wind developments in "Morocco."[131] The video opens with a low-angle shot of a sandy dune with dust blowing over it and a blindingly bright sun on the horizon, contrasting with the otherwise dark lighting. With regard to the soundscape of the Sahara, Siemens relies on the gothic. We hear a howling wind, emphasizing the visuals that suggest desolation—the desert wind is barbaric and threatening. Then comes the moment of colonial encounter, of the heroic missionary's brave confrontation with hostility. The mood changes as text appears on screen, signaling Siemens's entrance in this damned desert: "Siemens presents: Drawing the wind." Simultaneously, the light brightens, the wind's howl is replaced with an uplifting instrumental melody, and the camera cuts to a pan of beige desert and rock, finally resting on a row of mills, with blades circling—Siemens has successfully colonized and tamed the wind. The camera cuts to another low-angle shot, this time taken from the bottom of a mill, giving the latter a gigantesque, majestic look. This use of perspective emphasizes the enormity of the Siemens installations and shows off the white cylindrical curves. These, in turn, suggest futuristic technological prowess and elegant sophistication in design, which is further emphasized by its juxtaposition with flat, rocky scrub and spikey, sparse flora of the alleged terra nullius. Recalling the previous chapter, the aesthetic similarities with Atlas S.A.'s presentation of its fuel tanks in the Spanish colonial era is astounding: the oversized installations; the cylindrical, smooth surfaces suggesting order in contrast to the rough and "wild" desert surroundings, with its spiky, scrubby plants; and the futuristic white color reflecting bright sunshine, which, in a signal of power, would presumably force anyone in the presence of the fuel tank/wind turbine to squint. In both cases, wind imaginaries—both of the "hostile" wind itself and aeolian geomorphology in their references to the useless, windblown desert—are central to forming and communicating Atlas's and Siemens's colonial message. Siemens borrows and patches its discourse from the worn European colonial stereotypes with its recourse to ideas of *terra nullius*, the West's technological acumen in overcoming the climatological—mainly wind—barriers of the desert and the natives' lack of initiative in finding a way to put the "useless" and "hostile" desert and its "howling," "wild" winds to useful work. Adopting the hoary colonial discourses of white-man-as-savior, in the Siemens video opening, Siemens literally and metaphorically brings the light. Meanwhile, a voice-over claims, "To draw the wind, you must tame it." The wind is imagined as a barbaric, destructive, and wild agent that can only be harnessed (dominated and domesticated) by Siemens.

Such colonial discursive formations were also at work in the now-buried sorry saga of the DESERTEC project, to be led by the Desertec Industrial Initiative. The latter was a German-led private sector project, of which Siemens was a founding member, which eventually collapsed due to lack of investment.[132] The Desertec project envisaged generating solar and wind energy in the Sahara Desert, including occupied Western Sahara, and exporting it to the European energy grid. Plans paid little attention to the impact of such a project on locals, not least with regard to the vast amounts of water needed for the solar farms. Hamza Hamouchene interprets the project as yet another form of colonial natural resource exploitation, enriching Europe at the expense of North Africans.[133] Desertec Industrial Initiative's presentation at the 2010 Siemens Sustainability Conference opened with the following quotation from Frank Shuman, who in 1912 installed a parabolic trough collector in Egypt: "One thing I am sure of and that is that the human race must finally utilize direct sun power or revert to barbarism."[134] This colonialist nostalgia for overcoming barbarism, where barbarism was (and is) what fosters the self-definition and superiority of civilization, is reflected in Siemens's wider discourses on its work throughout Africa.[135] In an advertorial on its African portfolio, the company claims, "Electricity makes us human. Look anywhere in the world and you'll find that whatever people do that makes them human is powered by electricity" (emphasis in original).[136] This rather blatant civilizing mission discourse, and thinly veiled racism, underpins Siemens's articulation of its role in DESERTEC. A local engineer (wearing a Saharawi *melḥfa)* is quoted in an interview in Siemens's *Picture of the Future* magazine regarding Siemens wind projects in Morocco (including Foum El Oued, occupied Western Sahara) and the wider DESERTEC initiative. The article states, "Of course this doesn't mean [Morocco] plans to lease its [*sic*] desert to foreign power plant developers" and quotes the engineer, "We want to learn, and then some day establish our own industry for renewable energy and power plant components."[137] Rehashing white-savior narratives of altruism, Siemens positions itself as selfishly teaching the natives (ironically, the interviewee who symbolically takes the place of barbaric native to Siemens's missionary position here is actually simultaneously made a mouthpiece for Moroccan colonialism) how to take advantage of their unruly winds.

Conclusion

Siemens furthers Morocco's colonization of Western Sahara in a material as well as imaginary way. Its wind imaginaries of Western Sahara, of a barren and useless desert ravaged by a wild beast of a wind, are used in promotional materials to encourage a colonialist vision of Siemens's desert developments. The windblown wasteland can be made useful only with Siemens's know-how, advanced technologies, and environmental goodwill. Its tone is patronizing, implying that Moroccan natives would be incapable of developing such infrastructure themselves. And indeed, for Siemens, the natives are Moroccans, not Saharawis. Siemens, with its tamed desert wind, is an active partner in Moroccan settler colonialism.

Siemens's wind imaginaries drink from the earlier colonial wind imaginaries of the Spanish colonists on the one hand, and European and American shipwrecked sailors on the other. The tradition of weaponizing colonial wind imaginaries to justify a wider colonial project is maintained here by Siemens, which attempts, specifically, to justify its legally questionable energy developments in occupied Western Sahara. Wind imaginaries are a central component of colonial, renewable energy infrastructure.

CHAPTER 4

The *Tallīya*

Saharawi Perceptions of an Oppressive Energoregime

Before dawn, Mahjoub[1] sticks his head out of the upstairs window. He looks out across an abandoned building site where the dust is gathering in silvery swirls. If there is an *Asharqiya* blowing, it will be too dangerous to go out to sea. If there is a *Sāḥlīya*, there will be no point—there will be no fish. The sky is gray, and the air smells earthy and damp. Mahjoub smiles at the wind pushing the hair from his forehead. "Glory to God!" he says to himself. It's a *Tallīya*.[2]

The *Tallīya* blows from north to south. It's a northerly—it comes from the north. Like the Spanish colonizers. Like the Moroccan invaders. And now like the huge vessels that scour the seabed, deplete the fish stocks, and hog the licenses. Saharawi fishermen have it hard these days. Even for those who manage to pay for a sublet license from a regime crony, it is hard to compete with the European and Russian trawlers.[3] Nevertheless, Mahjoub is happy this morning. A *Tallīya* blowing means plenty of fish.

Mahjoub lives down the road from Mahmoud Alina, whom we met in chapter 2. Mahmoud is an old man now. He always wears his magnificent blue *darrāʿa*. He suffers from chronic illness and stays put with his family in the city. Mahmoud is sedentary in two senses: he lives in a stationary home, and he can barely walk due to illness. But he still identifies as a nomad. Precisely because he is nomad, Mahmoud explains, wind holds several meanings for him. "It's what brings good news and bad. Wind means life for us with its hard side and vice versa." "When the wind moves," says Mahmoud, "nomads can tell where it will carry the rain clouds." Nomads follow the signs of the wind and arrive—goats, camels, and all—to the place where rain is sure to fall. Then humans, animals, and plants enjoy the water. "All that, just from the wind."[4]

Unlike Mahjoub who was born in the coastal city of El Aaiún, Mahmoud is a man of the *bādīa*, and his relationship with the *Tallīya* is such. According to him, the *Tallīya* is usually pursued by storms that drench the desert in rain. When the *Tallīya* warns nomads that floods are on the way, they take time to prepare by way of postponing journeys, digging trenches around tents, and extending these trenches with canals to direct water away from living spaces. But, of course, such abundant rainfall "is good for Western Sahara in general."[5] However, the *Tallīya* can bring the *Ġalga*, an intense sandstorm that clouds all human visibility. Nomads can normally predict its arrival, and take necessary precautions, thanks to the wind, but if the signs are missed, the *Ġalga* can be a threat to life. "You can lose your children in it," warns Mahmoud.

Mahmoud lost a grown-up child. Not to the *Ġalga* though. He just disappeared. Rather, he *was* disappeared, taken during the 1999 Saharawi intifada, which demanded respect for human rights and raised socioeconomic grievances.[6] A human-shaped question mark. Moroccan rule in Western Sahara has been characterized, especially between 1975 and the late nineties, by mass slaughters, forced disappearances, and violent repression of Saharawis.[7] Even today though, occupied Western Sahara is still widely recognized as one of the worst places in the world for abuses of human rights such as civil liberties.[8] Such abuses are exacerbated by an energy system that is colonial in both imaginary and practical ways.

Just as European maritime empires harnessed the trade winds to further imperialism in the Americas and transatlantic slavery, Siemens, in its words as quoted in chapter 3, "harnesses" and "tames" the same "Atlantic trade winds." While the Spanish colonial research studies explored in chapter 2 highlighted the deep anxieties provoked by the Saharan winds for the colonizers, they also brought some nuance in the form of wind imaginaries that romanticized the desert and noticed the winds' creative and artful properties. The corporate discourse of colonial Atlas S.A., however, offered little subtilty or depth in its presentation of Saharan winds. Atlas's imaginations—delineated in chapter 2—of a hostile, destructive, and useless wind and windblown desert allowed it to mobilize all manner of civilizing mission discourse to aggrandize its activities. Siemens's use of wind imaginations is remarkably similar to those of Atlas. Both build on long-held ideas in the Global Northern mind of desert as wasteland and, specifically in the case of Western Sahara, a *wind-blown terra nullius*. This is little different to the imaginaries that shipwreck survivors constructed in their narratives some two hundred years before. My intention, in this chapter, is to further show how wind imaginaries shape energy systems by exploring the character of a (wind) energy system underpinned by colonial wind imaginaries. Specifically, I focus on the character of a modern energy system by seeking to understand how a colonized people experiences that energy system. I ask how the energy system in occupied Western Sahara, of which these colonial wind imaginaries are a constituent part, is experienced by Saharawis living with it. Saharawis perceive the energy system in occupied Western Sahara with hostility, I argue. Energy, including wind and solar energy, deserves special consideration among the multiple forms of natural resource exploitation that have contributed to the renewal of war in Western Sahara. This is because energy exploitation bolsters the occupation in a way other types of resource exploitation do not: energy infrastructure (in the form of cables, distribution lines, submarine cables, and so on) creates material links between Morocco, Western Sahara, and other countries beyond, and the (lack of) domestic energy affects Saharawi lives on an everyday basis.

Several scholars focusing on African cases argue that energy infrastructure is wielded by states to extend control over their populations, including by enlisting the latter in citizenship.[9] Tanja Winther and Harold Wilhite have argued that grid extension in rural Zanzibar provided the state with a mechanism for extending political control over the population.[10] Using the case of Mozambique, Marcus Power and Joshua Kirshner understand electrical technologies and infrastructure as "political terrain."[11] They argue that large scale electrical infrastructural developments are a way for the state to perform and narrate its own presence, as well as to enlist a state's citizens in projects of modernization and development. More specifically, Power and Kirshner argue that electrification allows the state to enroll its citizens as bill-paying customers of neoliberal capitalist development. Charlotte Lemanski's work on infrastructural citizenship finds that citizenship and identity are embedded in infrastructure and that public infrastructure is a local representative of the state.[12] Likewise, Idalina Baptista asserts that users' everyday interactions with electrical infrastructure elicit their critical positionality with regard to the state and thus to their condition as citizens.[13]

Other studies focusing on Global Southern contexts highlight the colonial implications of such power politics. For example, Charis Enns and Brock Bersaglio argue that large infrastructure projects in East Africa are reproducing colonial legacies.[14] Similarly, Power et al. contend that socially and spatially uneven energy infrastructural developments in Mozambique and South Africa are at risk of perpetuating a colonial electrical geography, given that they favor corporate and elite clients over households and communities.[15] Energy developments are also contributing to settler colonial conflicts in settings as diverse as Mexico and occupied Palestine.[16]

In this chapter, I build on this literature on the colonial and state-led politics of energy systems by focusing on how the colonized population experiences and perceives the energy system and how this in turn influences their positionality towards it. The concept of an energoregime is useful here. Dominic Boyer's concept of energopower is "an alternative genealogy of modern power" that "*rethinks political power through the twin analytics of electricity and fuel.*"[17] An energopolitical analysis goes beyond the study of energy politics to explore how energy, fuel, and electricity (are used to) govern populations and produce subjectivities. Douglas Rogers conceptualizes an energopolitical regime as a political system in which energy companies join state agencies in molding and transforming citizens.[18] I explore how Saharawis experience the Moroccan energopolitical regime in their territory and, following Timothy Mitchell's lead in showing how social movements take advantage of energy systems' vulnerabilities, demonstrate how the energopolitical regime furthers allegiance to a Saharawi nationalist conscience.[19] Michel Foucault argues that resistance is not exterior to power and that power and resistance are mutually constitutive.[20] Applied to an energopolitical regime, and as Mitchell has argued, the flows, cables, and pylons of electricity mediate political power and also opportunities for political resistance. Mitchell shows that fuel—specifically coal—offered political agency not due to subjective class consciousness or collective ideology but through the threat of sabotage and strike.[21] With the turn to oil as a fuel, which required fewer workers and whose infrastructure was not as vulnerable to sabotage, opportunities for workers' agency disappeared. Saharawis, victims of labor-market discrimination and a minority in their own country, likewise lack the tools of historic coal workers. Why and how, then, do they resist? I argue that oppressive colonial energopolitical regimes foster antagonistic subjectivities. In other words, an oppressive colonial energoregime will produce subjects hostile to itself. I first focus on how the energoregime antagonizes Saharawis at a local level through unequal energy distribution, orchestrated power cuts, and discrimination. Then I explore how antagonisms are furthered through the (threat of) violent oppression of Saharawis.

As well as arguing that Saharawis perceive the energy system as oppressive, I also give the energoregime in occupied Western Sahara the same qualification. But what conditions constitute oppression? Community psychologists find that oppression ensures "that the distribution of resources is unfairly allocated."[22] Oppression also necessitates the existence of a privilege group, argues Ann Cudd.[23] Another social group must privilege from the harm done to the oppressed group. Xavier Márquez suggests an authoritarian oppressive regime "is capable of heinous actions, including torturing and killing."[24] Milan

Zafirovski identifies an authoritarian, oppressive regime with "military-style discipline, brutality, and cruelty."[25] Demonstrating "brutality" or "cruelty" is a subjective task. For this reason, I make generous use, in this chapter, of the ethnographic method of thick description to give the reader a sense of the cruelty and climate of brutal violence faced by Saharawis.

Research for this chapter began in summer 2014, when Saleh and Abdelhay finally left me at Ali's[26] house after the detour to the Siemens wind factory detailed in chapter 3. This moment of arrival was when the usefulness of my gendered and raced disguise finally became apparent. Ali's house was permanently spied on. In fact, there was a spy sitting on a plastic chair outside the house minding Ali's business when I arrived. I could step out of the Land Rover, covered head to toe, and, like any Saharawi man, walk up to the front door and waltz inside without a second look from the spy. I did not step outside again, though, until it was time to leave the country. Ali and his siblings instructed me to stay away from the windows—who knows who might peer in. Although I made this trip long before beginning this book, and for purposes other than researching energy systems, that time at Ali's late grandmother's home was precious for research reasons as well as personal ones. I came to understand more clearly the impact of electricity and water cuts on Ali, his siblings, and the latter's children. I draw extensively on field notes and interviews from that trip in this chapter, as well as on interviews conducted by my colleagues Hamza Lakhal and Mahmoud Lemaadel.

Power Outages, Prohibitive Costs, and Energy Non-citizens

The power out goes after dusk. Zrug and his wife stumble around the room grabbing for the hands of their children, fearing an accident with the tea stove.[27] Zrug carefully descends the staircase, squeezing the rail with one hand and holding the baby in the other. He keeps matches, candles, and an oil lamp in a box near the front door. This oil lamp symbolizes safety for Zrug. Outside, his toddler clutches his mother's leg as other neighbors trickle out of their homes. They hold candles or oil lamps or the torchlight of smartphones. Zrug says that too often "we fall into a state of chaos [when there's a power outage after dark] that risks the lives of our kids."

Ninety minutes later, down the road in *Leḥšīša*, the bulb ticks and the light flickers in Safia's[28] living room. It momentarily reveals the brown and orange cushions that sit against the mint-green wall. Safia sighs as the light comes back entirely. She heads to the bathroom to fill water bottles. It will not be long before the water cuts out again. Her nephew Ali lies on his side on the carpet.

He tells me, "These cuts are usual in *Leḥšīša* and the other Saharawi suburbs, but you can bet the settlers still have their showers." He implies that the regular water and electricity cuts are a deliberate form of collective punishment. We can understand this punishment as a gendered one, as erratic electricity and water supplies add hugely to the domestic workload, which belongs almost exclusively to women.[29] Mahmoud, the elderly nomad, lives nearby. He too experiences frequent (monthly) power cuts, usually for six hours or more each time. "As a nomad, these don't affect me." He is used to life without electricity. But Mahmoud says the impact on his wife and the children that still live with him is considerable.

Mahmoud is seventy-eight now. He says that power cuts are "frequent." "They say they are due to problems in the grid, but we know that they sometimes cut the power on purpose when they want to bring secret things to the city, or when the young people protest in the streets." Fadel,[30] a young political activist, sheds light on what Mahmoud may mean by "secret things." Fadel says that blackouts coincide with Saharawi national days and Saharawi demonstrations, and "sometimes they cut [the power] if they bring more soldiers and arms from the airport to the desert, to the berm; they don't want people or activists to know how many arms, tanks, and soldiers are entering."[31] Hartan gives further insights: "When there is a homecoming of Saharawi political detainees, Moroccan-occupying authorities intentionally cut [the power] off in order to screw up the event. . . . I was personally able to see the suffering of media activists, and this was when we were trapped during popular demonstrations in conjunction with UN envoy Christopher Ross's visit to occupied El Aaiún. . . . I noticed how their camera batteries had run out so they couldn't monitor the violations."[32]

Scholarship on energy and colonialism highlights the links between electricity, unequal access, inequalities, and discrimination.[33] With regard to settler colonialism, Jabary Salamanca argues that Israel employs infrastructural violence in Gaza by controlling and sanctioning Palestinian access to key infrastructure.[34] Saharawis interpret a similar dynamic in occupied Western Sahara, alleging that their community's energy access is removed at Morocco's will and often linked to pro-independence and/or anti-occupation activities. Morocco employs energopower to curtail performances of Saharawi nationalism.

Interviewees insisted that the districts with the highest population of ethnic Saharawis—Maatalla, for example—are subject to more blackouts. Electricity access has long been a tool for race-based socio-spatial differentiation. In the American South, for example, governments developed electrical infrastructure that would institutionalize white supremacy.[35] This pattern was reflected in European colonies, where white settlers had better access to infrastructural services.[36] Since invading Western Sahara, Morocco has embarked on the settler-colonial policy of moving its own population into Western Sahara. Moroccan settlers now greatly outnumber Indigenous Saharawis. Comments such as Ali's above—"the settlers still have their showers"—reveal that infrastructure is understood as a tool, wielded by an oppressive colonial force, to differentiate the colonizers from "the natives."

Nguia is unusual among our interviewees in describing power cuts as infrequent. Like the aforementioned interviewees, though, she links frequency to politics. "At night, when the lights are out, violent confrontations erupt between Moroccan police and civilians. Therefore, power outages cause trouble for [the authorities]."[37] According to Nguia, then, the energy companies avoid blackouts to avoid civil unrest. Dadi's view contradicts Nguia's. He finds that power outages are an orchestrated consequence, rather than a cause, of political unrest: "[A blackout] happens for political reasons, for example, because of late-night demonstrations."[38] He recalls feeling "terrified" when, following demonstrations across El Aaiún upon the visit of the UN envoy, police began raiding homes under the dark cover of a power outage. It is worth highlighting that for Dadi, Nguia, and others, there is little distinction between the energy companies and the Moroccan regime security authorities. As Baptista points out, where a service provider is intimately associated with the state, the provider–user relationship is often understood as a proxy for state–society relations.[39] Saharawis talk of energy providers and the Moroccan state's security services as if they were one and the same: an energopolitical regime. Whether power outages happen to avoid confrontations, to quell political unrest, or to hide human rights abuses from international observers, the energopolitical regime is understood as the agent behind them.

Exactly how frequent is frequent when it comes to power cuts? Where Zain lives, power cuts were, until quite recently, a weekly or fortnightly occurrence, but they have become less frequent than this in recent years. Dadi describes cuts as monthly. Similarly, Najem finds they take place about ten times a year, usually for several hours. Zrug, who "frequently" experiences power cuts at his home, says they happen roughly every forty days. He adds that the district where his parents live, specifically Bou Craa Boulevard near the Fosbucraa mine, is known to experience a much greater frequency of cuts. For Zrug, the power cuts would not be "a big deal," but "the problem is there has been no official apology by the stakeholders or to send a prior notification to prepare ourselves and get our electronic equipment off to avoid disruption."[40]

Sixty-two-year-old Salka describes herself as an activist for peace. She puts power cuts in her home down to problems on the grid, or her inability to pay bills. She spends 50 to 60 percent of her monthly income on electricity bills. Meeting the energy and water bills is a great source of anxiety for her and, she says, "for all Saharawis." When she fails to pay, the electricity is cut.[41] She has also, on one occasion, had her meter removed, as have both Mohammed and Omar. Omar explains what happens if you fail to pay bills on time: "After sending warning letters, the water and electricity companies will cut you off and taxes you double what you owe if you want to be reconnected. Then, if you still can't pay, they will remove your electric meter."[42]

There was a general mistrust of suppliers on the part of interviewees. Several interviewees say they were being mischarged. Mahmoud says, "They sometimes send us invoices with the wrong amounts. In our house we don't have many appliances, so we know how much energy we use."[43] Najem and Najla share an anecdote of a time they spent away from home. Their usual monthly bill is 400–600 dirhams (Dh), yet they were sent a bill for 1,000 Dh for the month that the house was empty. They did not make a complaint. Najem explained, "We paid it because we can't do anything about this. Those who are in charge of bills in the electricity company are trying to steal money and there is no way to stop them."[44]

Without exception, all interviewees in houses connected to the grid found their bills expensive. Of those interviewees that had a regular income, on average, 30 percent of it was spent on electricity. All interviewees knew of families, notably those living in the slums of east El Aaiún, that did not have electricity in their homes. Several also mentioned families with low incomes that managed to "poach" electricity. Interviewees spoke of such families with no judgment. Eric Verdeil, in his work on illegal hookups in Beirut, conceptualizes similar poaching by Beirutis as a form of grid-embedded resistance at the "end of the line."[45] Verdeil finds it is an individual rather than collective form of dissent that is not necessarily connected to any wider political demands. Jonathan Silver notes a similar practice in poor areas of urban Accra. He conceptualizes these practices as "clandestine connections" and suggests they may be read as "the material articulation of a future with lower tariffs or even free energy for marginalized communities, an attempt by the neighborhood not just to envisage but to actually bring about a future of more equitable energy access."[46] Silver, then, allows the possibility to think of "illegal" hookups as collective resistance, not just individual dissent. Returning to Verdeil, and following his reading of Asef Bayat, "illegal" hookups show how the politics of Saharawi energy poverty are fought in the back alleyways of the poorest neighborhoods.[47] But not only there. Town squares, wide avenues, and the city's main thoroughfares have seen street protests concerning the high cost of electricity and water, bringing dissent into the public sphere. Zrug explains,

> We are in 2019 and in a few coming days, we will get into the year of 2020. I know many who do not have electricity at home. A lot of companies have launched big projects of energy power and, not far from these projects, people in El Aaiún are living without electricity. . . . There was a protest in Al Matara neighborhood concerning the energy and water outages and, by the way, despite the fact that we are just twenty-five miles from the sea and around us there are many groundwater points, the water cuts are more frequent than the electricity ones. Water is also energy.[48]

Dadi's aunt cannot afford electricity. She uses firewood to cook and to warm water for bathing. The high cost of electricity makes it an exclusive resource. But it excludes poor Saharawis from more than just reliable energy access. Lemanski argues that people's everyday access to public infrastructure affects their citizenship identities.[49] Failing infrastructure therefore constitutes a breakdown in state–citizen relationships. If we read energy infrastructure as a physical manifestation of (as Lemanski does with public infrastructure in the South African case) Morocco's citizenship model, then excluding Saharawis from energy access only fosters their further (as mentioned earlier, most scholars estimate that the vast majority of Saharawis are pro-independence) disassociation with the Moroccan state. While an energopolitical regime seeks to govern population and shape subjectivities, the oppressive nature of the energopolitical regime in Western Sahara produces subjects hostile to itself. In other words, the energopolitical regime has attempted to develop infrastructure in the southern provinces to, at least outwardly (in reality, infrastructural developments follow their traditional colonial purpose of facilitating capitalist wealth-gathering), serve Saharawis and integrate them into the wider Moroccan nation by way of offering development, modernization, and infrastructure, but has achieved the opposite. As energy infrastructure removes the border between Western Sahara and Morocco, it simultaneously bolsters the border between Morocco and the Saharawi Republic in Saharawi hearts and minds.

The Visibility of the Energopolitical Regime and Bodily Violence

Researchers working on perceptions of energy often comment that people do not think about energy except on the occasion of a power outage.[50] Energy is mystified. People consume it in great quantities with no awareness of its providence or how it reaches them. I join other scholars working on African contexts, such as Silver, in finding that, contrary to the body of research that finds energy to be invisible until it is absent in the Global North, in Western Sahara energy is "visible" to Saharawis in the sense that it is perceived as a dangerous mediator of oppressive politics whether materially present or absent.[51] Saharawis living in occupied Western Sahara interpret energy—its providence, its use, its cost in the widest sense of the word—in intensely political ways. A power outage is not needed to provoke thought or debate about energy. This is precisely because of the presence of an oppressive energopolitical regime in occupied Western Sahara. Below I highlight Saharawi *perceptions* of energy infrastructure: who Saharawis believe is behind energy infrastructure in occupied Western Sahara, the various political meanings that energy infrastructure constitutes for Saharawis, and the (physical) oppression and violence they (believe they will) face if they dare to publicly question the energopolitical regime. I continue with Zrug's story.

Zrug's case is exceptional because he is the only Saharawi (known to myself and my colleagues Hamza Lakhal and Mahmoud Lemaadel) that has benefited from a job in the energy industry. Zrug is passionate about looking after the environment. He is currently unemployed. Ideally, he would make a living by working in the field of sustainable development. He has so far been only partially successful in this aspiration. Zrug has worked for day wages for a solar thermal company, but before that, he worked for Geofizyka Kraków, a Polish company providing seismic data on hydrocarbons. He describes his experiences of working for energy companies as "exploitative."[52] He would work, on average, thirteen hours a day for these companies and says he was paid the equivalent of one euro an hour.

Before these jobs, Zrug had thought that energy used in occupied Western Sahara arrived from Morocco via transmission lines. But, through his work, he saw solar installations in Dchiera village, near El Aaiún, wind turbines near Dora and Tah villages, a wind farm outside Boujdour, and oil drilling near Bou Craa. Now he knows that Western Sahara generates and exports energy. And he also learned who profited from his exploitation. "Energy is managed and owned by multinational companies and individuals like the king of Morocco. . . . There are multiple companies that generate energy in Western Sahara while we are paying it from our wallets, and this is nonsense."[53]

Although Zrug is at pains to emphasize that these developments are illegal because Polisario does not consent to them, energy poverty is the most important factor for Zrug:

> Wind farms and so on are making the poor poorer and the rich richer. Green energy is being exported out of Western Sahara to other places in Africa and elsewhere. Although this is illegal because it is done by the Moroccan occupation, I feel proud as many elsewhere will be able to use electricity for lighting and other activities. They need electricity, just as I do. I am in favor of benefits for people everywhere, and I can compromise my rights for them to produce light for poor people, but under one condition: it has to be for free and not for sale.[54]

Zrug knows that his condition is not met. Even his own parents are not connected to the grid. They rely on candles for light. His father is over ninety years old, and Zrug worries for them. Ironically, they live between the energy developments at Dcheira village and the Siemens wind factory at Bou Craa phosphate mine.

Zrug's indignation at the unequal distribution of energy wealth pushed him to protest. A banner hangs from his house reading "Stop the plundering of Western Sahara's natural resources." He also recently participated in a fifty-two-day sit-in in Mecca Street, El Aaiún, demanding jobs for Saharawis, and another lasting twenty-nine days at the Guerguerat buffer zone (Saharawi civilians had intermittently been setting up roadblocks and other forms of protest at Guerguerat since 2016, well before the November 2020 outbreak of war, which occurred at Guerguerat roadblock). The latter protested illegal exploitation of Saharawis' natural resources. Zrug has since ceased his protesting activities. After Guerguerat, he managed to hustle some part-time work acting as a bus driver. But, he says, Moroccan authorities got him sacked. He is also banned from traveling outside of Aaiún. Zrug fears for the well-being of his children, wife, and parents: "No one will feed them if [the authorities] take me."[55]

No interviewees had ever been invited to take part in any consultation process on energy developments. All interviewees were likewise unanimous in their view that these energy developments held zero benefits for Saharawis. As Nguia explains, "We [Saharawis] are regarded as third- or fourth-class citizens, and there is therefore no need to invite us to take part in a consultation."[56] All interviewees also agreed that the benefits of natural resource exploitation go to Moroccans—mostly wealthy settlers and ex-high-ranking military personnel and government members—and a small minority of wealthy Saharawi families with close ties to the Moroccan authorities. This is reflected in research that focuses on other areas of natural resource exploitation.[57] Energy infrastructure is understood as funding the patronage networks that maintain elite allegiance to the Moroccan occupation and king.

Interviewees expressed positive views toward renewable energy in general, but none had positive opinions of the wind and solar factories in occupied Western Sahara. Who do Saharawis believe is responsible for the developments? Nguia believes it was a "foreign" company called Siemens, which has "no humanity." She adds, "The occupying power is letting other countries invest here as a way of getting them to recognize Moroccan sovereignty over Western Sahara."[58] Najem expects that most developments would be the result of partnerships between French companies and the Moroccan king.[59]

Hartan has heard of Kosmos Energy.[60] Dadi, too, has heard that Kosmos is active in the region, as well as San Leon. He opines, "These companies contribute to Moroccan colonization and endlessly support its presence."[61] Dadi, Hartan, and Ahmed collaborated to create news articles about an event held in El Aaiún on the Noor 1 (Dchiera, near El Aaiún) solar project, to which Saharawis were not invited. In retribution, Ahmed, who is a university student in Morocco, had his study and travel grants pulled, and Moroccan authorities also canceled social benefits received by his relatives.[62]

Salka knows about solar energy projects in El Aaiún port and Tarfaya, but she is not aware of the names of any companies involved in these developments. She comments, "All profits go to the Moroccan occupation and foreign companies."[63] She has never publicly protested these energy developments because she is afraid of violent repression. Omar and Mohammed, like Nguia, voice their desire to demonstrate against the energy companies present in their country, but they are too afraid to do so.

Mahjoub is the fisherman with whom I opened this chapter. He does not identify as an activist. He knows of solar and wind farms at Tarfaya and El Aaiún port. He believes them to be owned by the king but knows that Siemens is also involved. Does Mahjoub like the look of wind turbines? What do they represent for him aesthetically? Mahjoub giggles and remarks, "They do not represent anything but a scene of the wind of your land being illegally exploited by the invaders with no benefits for the people."[64]

As for Fadel, he comments, "I know Siemens and the Moroccan royal holding and others. They support this occupation and give Morocco reasons to continue the occupation."[65] He says he has participated in a peaceful protest against such energy companies and was beaten by Moroccan authorities for doing so. Mahmoud does not know what green energy is. He has seen a wind turbine though. He describes it as "a machine or tool made to steal our resources and say we benefit from it."[66]

Zain is a student. He does not identify as an activist. He knows of wind factories at Boujdour, Tarfaya, and Dakhla and of a solar thermal project at

Dchiera. When asked if he knows who manages these developments, he replies in a whisper, "Can I say the word occupation?"[67] When reassured, he goes on to talk of four Saharawi families in Tarfaya (Tarfaya is in Morocco proper, but was historically roamed by Saharawi nomads) whose land was confiscated, with no compensation, to make way for the Siemens developments there. As discussed in chapter 3, Western Sahara Resource Watch has reported on one such family in Tarfaya, as well as on the case of an El Aaiún family who lost their home (and whose members were beaten and arrested when they refused to leave the house) to make way for an Alstom-constructed power line.[68]

Salem does not see himself as an activist. He feels energy developments in occupied Western Sahara cause his people suffering. He says that "the consultation about such projects is always only with well-known individuals and they negotiate deals. Ordinary people are not consulted."[69] In the past, Salem has emailed foreign energy companies to urge them to divest from his country. He received no replies. He also attended a demonstration against the solar developments but was beaten there by police.

Interviewees' views of the developments and the developers were uniformly negative because they "fuel Moroccan colonialism and occupation." That several interviewees named the Moroccan king as behind energy developments is significant given that, following Baptista, where infrastructure is associated with the state, it is often understood as a proxy for state–society relations.[70] That developers were also identified as *foreign* multinationals emphasizes the perceived coloniality of the energoregime. I argued in the former section that energy supply issues and prohibitively high energy costs foster an othering of Saharawis and thereby a rejection, on their part, of Moroccan citizenship. In this section, we see that energy developments are met with disdain largely on Saharawi nationalist grounds: the developers and developments are *colonial*, and they further the *occupation*. The oppressive energopolitical regime fuels a preexisting Saharawi identity and, with it, resistance. Beatings, job losses, benefits cuts, threats to relatives, and travel bans are the lot of those interviewees who have occasionally denounced the hated foreign energy developers. This violence is indicative of a situation of climate necropolitics: it is acceptable that Saharawis face violence in the name of climate change mitigation.[71] So far, I have cited interviews with Saharawis that did not consider themselves activists, or who were low-profile activists, in order to gage the views of "ordinary" Saharawis. If we turn, though, to the retributions suffered by those who dedicate their lives to struggling against resource plunder, we get a sense of the cruelty, degradation, and brutal violence that allow me to qualify (according to the definition explored at the outset of this chapter) the energopolitical regime in Western Sahara as one of authoritarian oppression.

Sultana Khaya, president of the Saharawi League for Human Rights and Natural Resources, has led several campaigns against specific energy companies, as well as for respect for Saharawi human rights. Here are a few facts about Sultana: She was born in Boujdour in 1980. Her most cherished place in the world is Tiris, Western Sahara, but she has never been there. Her role models are El Wali Mustapha Sayed (founder of Polisario) and Nelson Mandela, in that order. Her favorite musician is the late Mariam Hassan, who "moves [her] spirit," and her preferred writers are Malainin and Hamza Lakhal. Sultana has only one eye. Police beat the other one out of her head on the campus of Cadi Ayyad University, Marrakesh, in 2007. She was studying there for a degree in French at the time. Sultana, her sister Luara, and her mother have been under house arrest since November 19, 2020, which is, as I write, one year and two months ago, to the day.

According to Amnesty International, Sultana's house arrest is part of a wider Moroccan state crackdown—started at the outbreak of the new war—against Saharawi activists "away from the attention of international media."[72] Authorities have not provided Sultana with a reason for her detention or any legal basis for it. Over the time of the arrest, police have regularly stormed Sultana's home. She sometimes sends me videos on WhatsApp of the state of the house after their visits. Bundles of clothing and blankets are scattered across the floor. These fabrics mingle with triangles of broken stained-blue glass, turned-over mattresses, and toppled furniture. Cupboard doors gape open. There is nothing left inside them. One day, in January 2022, I have a video call with Sultana. She opens her mouth to show me the newly missing teeth—the gaps there that gape like those cupboard doors. But that is not the worst thing the intruders have done—Sultana starts to run out of words. Reports from international human rights NGOs on these unspeakable incidents make for a catalogue of horrors: "Dozens of masked members of the security forces raided Sultana Khaya's house, assaulted and attempted to rape her, and raped her sister," reads an Amnesty International report.[73] The Irish NGO Frontline Defenders circulates a report following an earlier police visit: "At least 40 plain-clothes Moroccan security agents raided the home of woman human rights defender Sultana Khaya, forcibly entering through the roof. When they reached Sultana Khaya's bed, the police officers covered her nose and mouth with a rag, which was soaked in a liquid chemical, that made her lose her ability to smell afterwards."[74] Another report details how police went for Sultana's remaining eye by beating her face with a rock.[75] As the months go by, the "Sultana Khaya" dossiers of Amnesty International and similar organizations get thicker and thicker. But nothing changes. "We need witnesses, human rights observers," Sultana insists, "and Saharawis need cameras—police keep stealing them to stop us showing the world what is happening here."[76] Dunlap has argued that displacing colonial, capitalist "renewable" industries to Global Southern margins would facilitate the ability to use extreme forms of violence against those who protest.[77] Sultana's case proves him right.

Morocco's oppressions of Sultana over the last fourteen months have also involved cutting off her energy supply. The darkness accentuates the horror of police raids on the home. It also makes it harder for Sultana to document and communicate the attacks that she, her mother, and her sister are suffering. No electricity means that there is no Wi-Fi, no laptop to work on (police stole her computer in any case), and difficulties in recharging her mobile phone. Also, there is "no power for the fridge." Sultana explains, "My mother is eighty-four and suffers from diabetes. As you'll know, people with diabetes need insulin injections, and these should be kept cool in the fridge."[78] At the start of her house arrest, police cut the power line into Sultana's house. Then, on April 8, 2021, they removed the electricity meter. The Khaya family has had no power since then. "We tried to get a solar panel on the roof to charge a battery, but no one is allowed in or out the house to help us set it up."[79] While occasional power outages are a form of collective punishment of the Saharawi community in general terms, high-profile activists such as Sultana suffer a permanent blackout.

The president of the only other Saharawi NGO with a specific and primary focus on preventing natural resource exploitation (the Committee for the Protection of Natural Resources in Western Sahara [CSPRON]) is Sidahmed Lemjeyid. To understand his fate, we should look back ten years.

It is December 25, 2010. San Leon Energy's executives have been overseeing oil exploration surveys in occupied Western Sahara all year. Now they enjoy a very white Christmas as the so-called Big Freeze of snowy weather falls on Dublin. Down in Western Sahara, the weather is a mild 20°C. There is a northerly wind, and clouds pass above. Police pick up Sidahmed Lemjeyid and Izana Amidan together. Izana is a high-profile human rights activist. Sidahmed, before his arrest, had dedicated all his time to his role as president of the CSPRON. This is an unofficial, unpaid role because Saharawi organizations that work to denounce activities such as San Leon's oil exploration surveys or Siemens's wind factories are not allowed to register or officially exist in any way.

I write elsewhere about what happened to Izana after her arrest.[80] In brief, police put her in a cell in a secret detention center, suspend her from a horizontal bar by way of ropes tied to her wrists and ankles, and gang rape her for three days before releasing her without charge. Three days. Sidahmed is also put in a cell—he does not know where it is—and is tortured, raped, and made to sign a confession that he has not read. The next day, his case is referred to Rabat's military court.[81] He spends the next three years being tortured in cells in El Aaiún, Dakhla, and Sale. A report into the torture suffered by Sidahmed and his compatriots makes grim reading: electric shocks, deprivation of food and water, being suspended in the "Vitruvian man" and "roast chicken" positions and exposed to beatings and sexual violence in these positions.[82]

In February 2013, Sidahmed stands up in court. He faces trial alongside twenty-three of his compatriots on trumped-up charges linked to the 2010 Gdeim Izik Saharawi protest, which was a month-long encampment of twenty thousand Saharawis demanding jobs, independence, and an end to natural resource plunder. Sidahmed wears a blue *darrāʿa* and hangs a green turban around his neck. Scars are visible on his hands and head. Someone is taking photographs. He looks at the camera and makes the "*V* for Victory" sign. They all do. The defendants give their statements. Sidahmed quotes, in his defense, the 2002 UN Legal Opinion on oil exploration activities in Western Sahara: "If further exploration and exploitation activities were to proceed in disregard of the interests and wishes of the people of Western Sahara, they would be in violation of the principles of international law applicable to mineral resource activities in Non-Self-Governing Territories."[83]

On sentencing day, there are power outages in El Aaiún. Saharawis say the Moroccan authorities have arranged for the blackout. Without electricity, it will take longer for the Saharawi community to learn the fates of their Gdeim Izik heroes, and it will be harder to coordinate large demonstrations. Slowly, the news perforates the El Aaiún darkness: Sidahmed is sentenced to life imprisonment. I read more of the aforementioned report: "forced ingestion of faeces, urine and cockroaches," "removal of finger and toenails," "exposure to cold," "taken to the cells of other inmates who are told to sexually assault him with objects that prison staff give them."[84]

Three years since, there is a small flame of hope in Sidahmed's lonely cell. He and the rest of the Gdeim Izik group are brought before Sale Civilian Appeals Court. In the courtroom, they stand in a glass cage. The trial lasts seven months. Some days, the defendants come in bloody—even wrapped in blankets. They claim to have been tortured on court premises, including in the presence of the investigative judge. On days that they are unable to stand up, they must lie in an adjacent room, where they cannot hear proceedings. On

March 21, 2017, Sidahmed gives his declaration. He mentions a phrase that he has been forced to utter while being tortured. It is a phrase that mocks his work but also mirrors the discourse of the occupiers and their corporate partners. "Morocco is good, it develops its southern provinces."[85]

On the last day, the court delivers the sentence within ten minutes.[86] Sidahmed is sentenced to life imprisonment.

Conclusion

El Aaiún, August 2014. Every night, after dark, the entire family relocates to the roof to enjoy the *Sāḥlīya* blowing off the sea. Small clouds glide eastward—orange. "The clouds are swimming," observes Ali. The roof terrace is bare and dusty, with a ruined brick pen in one corner. Ali's elder brother Youssef[87] built the enclosure for their grandmother—who brought them up after their mother's death—to shelter her goats. Youssef, a human rights and anti-plunder activist, was twice forcibly disappeared by the Moroccan state and eventually fled across the military wall, fearing for his life. Somehow he made it through the seven-million-strong minefield adjacent to the wall without losing a limb. He now campaigns against Siemens and other companies from the camps. He can never come back here.

Years since Youssef's departure, after the death of Ali and Youssef's grandmother, their aunts Safia[88] and Maimina[89] disagreed over what to do about the roof. Maimina wished to demolish the unused pen, tile the walls, clear the floor, and create a roof garden with potted shrubs and flowers. Safia shared a special bond with Youssef (I notice how her eyes fill whenever she speaks of him). She wished for the terrace to stay exactly as it was the day he left. Safia won the argument. The roof remained a time capsule.

Out in the time capsule, we use Ali's mobile phone for music. Bruce Springsteen, Mariam Hassan, Tracy Chapman, Tinariwen—Ali plays Sam Cooke's "A Change Is Gonna Come" two nights running. "It's like he's singing about the Saharawis. Listen! The tent, the river, the discrimination . . ." Some nights, Ali's nephew entertains us with his poetry. He is only eleven but already performs his work standing tall, without shyness. He speaks of winds, plunder, energy, sand, independence, the *bādīa*, and freedom.

In Western Sahara, energy justice is tangled up with self-determination. Siemens is perceived (and widely despised) as an agent of occupation and colonialism. Siemens imagines and presents itself as a benevolent European colonist taming wild, destructive winds for the benefit of implicitly incapable "natives." The resulting energoregime of which Siemens forms a part is equally colonist and equally preoccupied with "taming" the natives. Wind imaginaries are a constituent part of (wind) energy systems, even when such a system is fueled by fossil sources as well as sun and wind. Colonial wind imaginaries underpin colonial energy systems, and both are embedded in wider colonial politics. As I have shown in this chapter, though, such energy systems also breed opposition. The oppressive nature of the energopolitical regime produces resistance among Saharawi "natives." Nevertheless, for a Saharawi in the occupied territory, to contradict the logics of the energopolitical regime by protesting the foreign energy companies is to risk economic dispossession and bodily violence. For repeat offenders like Sidahmed Lemjeyid, it also means life imprisonment. In the prison cells, the energopolitical regime's electricity flows through the tortured body.

Part Three

The *Bādīa*

CHAPTER 5

The *Gallāba*

Windblown Desertscapes and the Friendship Generation

Tifariti, liberated Western Sahara. Windblown particles of dust and sand absorb and reflect light, allowing air to take on color. The sun's light cools and then warms throughout the day. It is filtered by clouds and acacia leaves and then reflected by wind-polished rock. The changes in light in turn alter the air's colors. The sky is perfect blue—then later milk and rose-pink. Western Sahara's *bādīa* could exhaust any artist's palette thanks—in part—to the work of its aeolian elements.

And here I am in the *bādīa*, with Andrea, Kari, Maimina, and our other new friends. It is 2006, I am a young student, and I feel so determined. Clouds are white boats above. You can hear the songbirds for miles. The sun and breeze hug us.

Wind and sun awaken our emotions, and, likewise, we interpret the weather through our emotions. The perfect weather during my first trip to liberated Western Sahara in February 2006 further elevated my happy state. Entire communities also share common climatic sensitivities. Rain in European and North American popular culture is often associated with sadness. For Saharawis, the various winds of their lost lands each evoke specific emotions. The Friendship Generation, in particular, finds hordes of emotive imagery in the windblown desert. In this chapter, I focus on how this group of Saharawi writers—the Friendship Generation—employ Saharan winds and windblown landforms to emotively present the *bādīa*, the beloved Saharawi nomads' heartlands, to a foreign, Spanish-speaking audience. This allows me to explore an alternative to the colonial wind imaginaries discussed in the preceding chapters of this book and, likewise, allows me to discuss in forthcoming chapters how an Indigenous wind imaginary might inspire a very different energy system to the problematic ones of Spanish Sahara and occupied Western Sahara.

The Saharawi Friendship Generation, constituted as such in 2005, is a group of Saharawi poets that, for the most part, share common biographical elements: they were born in Western Sahara, fled as children to the Saharawi refugee camps in Algeria, received scholarships to attend secondary school and university in Cuba, returned to the camps as graduates, and then moved to Spain in search of livelihoods that the refugee camps could not provide. Having spent over a decade of their formative years in Cuba without an opportunity to visit, or in the first years even telephone, their *Ḥassānīya*-speaking families back in North Africa, the generation writes in Spanish, which has become a surrogate mother tongue for them. Spanish also makes the generation's work accessible to audiences in Spain and Latin America, where the writers hope to find international allies and sympathizers to their cause.[1]

I focus here on two poets—Limam Boisha and Fatma Galia Mohammed Salem—who are members of the Friendship Generation. I argue that these writers employ what I call aeolian aesthetics. I borrow the first term from Howe and Boyer's conceptualization of *aeolian politics*, which highlight the manifold effects, negative as well as positive, of wind power. A Saharawi aeolian aesthetic is a particular wind-infused form of artistic expression inspired by the traditional way of life of Saharawi nomads and their relationship with wind and windblown desertscapes.[2] Aeolian aesthetics, I argue, allow Saharawi poets to claim their desert heartlands based on love and knowledge, to challenge dominant Western imaginaries of desert, and to counter hegemonic understandings of energy and their colonial implications. The artistic texts explored in this chapter undermine the logic of wind energy developers' colonial narratives explored in part 2. They do so by drawing on the properties of wind and aeolian geology aesthetically and by making apparent, explicitly for a foreign audience, how wind shapes and enables the Saharawi people's very way of life.

Because wind energy developments in Western Sahara emerge from, and maintain, the capitalist, colonial, oil-dominated world order, in this chapter I engage with petrocultural studies. Petroculture is the global culture of the modern era born of a colonial, capitalist, fossil-fueled energy system with all its political, social, economic, and environmental implications. Scholars of petroculture have concerned themselves with how oil has shaped modern culture and society, largely by reading energy into literature and visual culture. There is an aspiration among such scholars to contribute to a successful energy transition: only by unraveling and understanding the logics of oil-fueled culture and our attachments to them can we begin to reimagine and reinvent our anticipated lives after oil. Within petroculture as a field, scholarship to date has understandably focused on oil. While petroculture scholars have speculated that solar and wind literature may emerge in the future, post energy transition, I wish to build on the small emerging body of work by scholars such as Rhys Williams and Cymene Howe, who show that solar literature, or at least the use of aeolian aesthetics, already exists in the here and now.[3] In turn then, a discussion of aeolian aesthetics can yield positive outcomes beyond the Saharawi case. If knowing how oil has shaped culture is essential for ensuring a true energy transition, then surely also is knowledge of the cultural impact of energy sources to which we wish to transition. To transition away from a colonial energy system at a cultural level is to rethink how we understand energy. Imre Szeman and his coinvestigators have called for a project of "indigenizing energy," which would seek to understand the philosophies of Indigenous energy cultures and their implications for a global energy transition.[4] I follow that call by focusing on conceptions of wind, of wind energy, and, relatedly, of desert ecologies that emerge from Saharawi poetry.

Children of the Wind

Graeme Macdonald argues that "petroleum culture is enacted wherever there is a detectable reliance (conscious or otherwise) on fossil energy."[5] Since all modern cultural production emerges from an age of global petroculture, argues Macdonald, one can read oil into all world literature.[6] If this is the case, could Saharawis be an exception? I do not mean that Saharawis live outside of petroculture. Even those staying in the remote camps of Algeria are connected to the rest of the world by mobile phones, can hitch a lift in a Polisario Land Rover, and drink water from plastic bottles. Their exiled lives are furnished with petrocommodities, even if the material wealth and consumerism of petroculture is impossible for them. Besides, Saharawis' very exile and occupation are sustained by petrocultural energy systems. Nevertheless, lives before petroculture descended on Western Sahara are still in living memory. At least until the sixties, Spanish colonialists avoided interference with the "hostile" Saharawi tribes and largely stuck to the lucrative opportunities of Western Sahara's coastline. Furthermore, the Saharawi state-in-exile today attempts to conserve traditional culture by fostering the breeding of camels in the camps and clearing mines in "liberated" Western Sahara to allow for nomadic camel pastoralism there. Several middle-aged and older Saharawis knew nothing but the desert nomad's life until the Moroccan invasion. It is the memory of this life, and this life that is still practiced on the margins of petroculture in the "liberated" territories, that shapes the Saharawi poetry analyzed in this chapter. As I discuss below, the wind, the sounds it transmits, the water it carries as cargo, the trails it leaves in the sand, and the geological formations it creates have been central to the possibility of Saharawis' nomadic life for millennia. Saharawi culture is windblown. Just as petrocultural societies are dominated by oil aesthetics and the cultural production of Caribbean societies, which are built around imperial sugar plantations, is marked by saccharine stylistic tendencies, Saharawi cultural production is characterized by an aeolian aesthetic.[7]

Wind is elusive to the eye. We can see the wind only in the movements of the things it carries or in the visible traces of the material it has sculpted, destroyed, created, or left behind. Visual aeolian aesthetics therefore rely on images of the windblown. In Friendship Generation poetry, such imagery is drawn from aeolian geomorphology—that is, desertscapes created by the wind. Saharawi culture is constituted by these desertscapes, both those that are perceived as Saharawi, and—since all identities are delineated by what is other to them—those that are not.

Boisha evokes the windblown Algerian Hamada, which is the location of the Saharawi refugee camps, to comment on the Saharawis' victimization. "Would the Hamada exist if they had not tried to bury us there? / Would it exist if they had not told us so?" asks Boisha in his poem "Say That You Haven't Told Me," the title of which comes from a *Ḥassānīya* phrase that Saharawis use to avert bad luck when someone is about to share news.[8] Boisha's question highlights that places are not neutral entities, but instead socially, politically, and culturally constructed ones, and simultaneously reveals how the Algerian Hamada is imagined in Saharawi society: an infertile, bleak, and insufferably hot plain. But to say that places are socially constructed does not mean that they are materially inexistent.[9] *Hamada* is a term for a particular type of desert landscape characterized by rocky plains where almost all sand has been removed by an aeolian process known as *deflation*. Geologically, the Algerian Hamada would not exist without a huge and sustained wind power. The intense, turbulent—that is, characterized by chaotic changes in pressure and speed—action of the wind carries the sand away. Just a stony plateau of gravel and bare rock is left. Vegetal life is almost completely absent, and thus, nomadic camel pastoralism, the basis of Saharawis' cultural identity, is not easily possible. For the Saharawi imagined nation, the Hamada is—materially as well as aesthetically—apocalyptic.

Would the Hamada exist if they had not tried to bury us there?
Would it exist if they had not told us so?
Would it exist for others before we knew of its geography?
And for those who know it not?

Would the Sahara exist without the envious memory of the wind, without the fire's scars, the pastures' freedom, the acacias' shade?

Without the wall that separates our flesh, without the threads that sow death, our blood, would we exist?[10]

¿Existiría la hamada si no nos hubieran intentado enterrar en ella?
¿Existiría si no nos hubieran dicho que existía?
¿Existiría para otros antes de sabernos parte de su geografía?
¿Y para los que lo ignoran?

¿Existiría el Sahara sin la envidia de la memoria del viento, sin las señales del fuego,
la libertad de los pastos, la sombra de las acacias?

Sin el muro que separa nuestra carne, sin los hilos que siembran la muerte, sangre nuestra, ¿existiríamos?

The first line of "Say That You Haven't Told Me" nods to the international politics that have shaped this negative imagination of the Hamada. Saharawis have been "bur[ied]" there—trapped, their nationalist aspirations thwarted, and denied dignified lives as individuals, by an undetermined third person (this could be the Moroccan state, Spain, and/or the wider international community). The ecological and visual hostility that the Hamada represents for Saharawis is reflected in the tone of Boisha's poem. The imperative "Say" (*Di*) of its title and the seemingly relentless, vertically stacked rhetorical questions suggest the angry tirade of a wronged person against a culprit. The powerful, destructive potential of a scowling and scouring wind capable of creating a geological Hamada is reflected in the violence of the final line, with its arresting images of "flesh," "death," and "our blood." The latter possessive pronoun indicates Saharawi victimhood.

Yet in this poetic picture of the wind-created, or wind-destroyed, Hamada, Boisha manages to introduce a simultaneous but strikingly different symbolic role for wind. "Would the Sahara exist without the envious memory of the wind, without the fire's scars, the pastures' freedom, the acacias' shade?" Here blows the wind of the *bādīa*. *Bādīa* is the *Ḥassānīya* term for the beloved desert heartlands of Western Sahara. As Mahan Ellison points out, the *bādīa*, in Saharawi poetry, articulates Saharawi notions of identity, community, and home.[11] With its juxtaposition next to the "fire's scars," wind here symbolizes, on the one hand, resistance. On the other hand, wind is freedom. This is evident metaphorically through Boisha's evocation of Western Sahara's *bādīa* landscape and crudely through the reference to the nomadic life pursuing "pastures": nomads knew no walls and roamed freely. Saharawis today talk of the *bādīa*'s flowering and green havens after rains and lovingly recall its *āḫīām* (traditional Saharawi tents), its camels, its acacias, and its breezes. As ethnobotanist and anthropologist Gabriele Volpato explains, in Saharawi society, health, pride, dignity, and freedom are seen as linked to the *bādīa*.[12] Ellison likewise highlights that these Saharawi desert heartlands are an "affective space" and a site of nurture for the Saharawi Friendship Generation.[13] The *bādīa* is all that the Hamada is not. But it, too, exists as a product of wind power. The desert is a museum of aeolian geomorphology—that is, landforms shaped by the wind. While the aeolian process of deflation has created the rocky plains of the Hamada (and, of course, landforms likewise play a role in creating the winds), abrasion, the process through which wind-driven and windborne sand particles erode the earth's surface, has sculpted the rock formations that characterize the *bādīa* landscape. These rocky wind creations, called *gallāba*, or *galb* in the singular, make the very way of life of Saharawis possible. In his poem named after these formations, Boisha explains the sociocultural significance of *gallāba*.

A traveller asks me
what is the meaning of Galb.
I tell him, for example,
that Miyek is a blemish
in the belly of the earth.

That Ziza, for example,
is the Berber for breast,
and that the wing of a dune
can touch the sea of the sky.

I say, for example,
that in the high peaks
of prismatic dawns

there is much sleeping life
rubbing its skin.

That in the transient stone
there are stationary craters,
islands emerging
from an ocean of nothing.

For example, a Galb might be
the name of a girl
engraved
into the eyelashes of a cave.

As Tiris is the navel of the
Sahara,
Galb is a heart,
a heart of stone. [14]

Me pregunta un viajero
qué significa galb.
Digo yo, por ejemplo,
que Miyek es un lunar
en el vientre de la tierra.

Que Ziza, por ejemplo,
es pecho en lengua bereber,
y que el ala de una duna
puede tocar el mar del cielo.

Digo yo, por ejemplo,
que en los altos picos
de prismáticos amaneceres
—frotando su piel—
hay mucha vida dormida.

Que en la piedra pasajera
hay platillos estacionados,
islas que emergen
desde el océano de la nada.

Un galb puede ser, por ejemplo,
el nombre de una muchacha
esculpida
entre las pestañas de una cueva.

Como Tiris es el ombligo del
Sahara,
Galb es un corazón,
corazón de piedra.

The poem, in which Western Sahara is personified as a longed-for and desired woman, relies on the double meaning of *galb*. One meaning for *galb* is heart. It is also the *Ḥassānīya* word for a particular type of desert landmark. According to traditional *Ḥassānīya* epic poetry, which had (and has) a pedagogical use for transmitting ethnobotanical and geographic knowledge between the generations,[15] there are 365 *gallāba* in Western Sahara, each with its own name.[16] British and Spanish readers might have various words, shaped by European historical cultural understandings and ecological knowledge, to classify the physical manifestations that are *gallāba*—for example, mountains, caves, or rock formations. But for Saharawis, *gallāba* are the markers—physically resembling hearts—forged by the desert wind to enable their nomadic culture. Saharawi nomadic life is driven by the need to know where to find water and the best forage for their milk-providing, life-sustaining camels.[17] As Volpato has highlighted, Saharawis have developed strategies of temporal and spatial mobility to ensure their ability to locate the best pasture areas at any given time.[18] The wind-forged *gallāba* have made these mobility strategies possible. They are vital for the nomads' navigation through the desert as well as offering the protective lure of shelter and shade from the sun and wind.

Boisha's use of aeolian geomorphology as a source of structural inspiration and imagery serves to claim Western Sahara for Saharawis due to their knowledge and membership of the desert ecology. In authoring "Galb," Boisha takes the place of the wind in sculpting out stanzas that, like *gallāba*, guide us through Western Sahara. Mirroring the physical diversity of *gallāba*, each stanza has its own formal particularities: some are four lines long, some five; there is no regular rhyme scheme. But each stanza is similar in line length and content. All begin with a toponym, which is likened to an enticing part of a lover's body, and each ends with a lyrical flourish that employs imagery of lightness, flight, or water, suggesting freedom and fertility. Each stanza, each *galb*, allows us to navigate through the poem as if the entity of the poem was itself a metaphor for nomadic Saharawis' Western Sahara. The poem's formal features—short lines, the structuring of the poem into stanzas, and the abundant use of commas and full stops—paces the poem and ensures it progresses without unnecessary haste, as if to give the reader/listener time to fully pay attention to the desert surroundings. Thus, we are encouraged to mimic, or at least reflect on, a nomad's close relationship with all the desert elements. Boisha is nodding at Saharawis' intimate and complex knowledge of the botany, ecology, and meteorology of the *bādīa*, and of well locations, distances, and trajectories for crossing the desert.[19] Knowledge of place gives Saharawis the right to the *bādīa*, Boisha's poem suggests.

But Boisha also claims Western Sahara for his people based on love. He does so by using aeolian aesthetics to evoke regions with special cultural meaning in the Saharawi communal imaginary. As we work vertically down the poem, each stanza/*galb* leads us gradually southward, as if following the Saharawi nomads—guided by the *gallāba*—on a migration to Tiris. Tiris is considered the most beautiful region of Western Sahara's *bādīa*. It is located near the Mauritanian border, and indeed crosses over into Mauritania. The final stanza, or southernmost *galb*, is dedicated to Tiris. It begins, "As Tiris is the navel of the / Sahara." Leaving "Sahara" to stand alone gives the separateness, the exile from this wonderful place, an added weight and aura. The very mention of Tiris, the most cherished part of the Western Sahara *bādīa* heartlands, will produce an emotional response in a Saharawi audience. It is fitting, then, that Boisha ends with an image of a steadfast heart: "Galb is a heart / a heart of stone." The repetition of "heart," separated by a line break and comma, makes sure that the poem sits down and lingers on the imagery of a heart/*galb*. One is left, at the close of the poem, with all the poignant emotion of longing for, and missing, someone or somewhere. Yet the final line is also ambiguous. The simile suggests both the permanence and intransience of Saharawis' love for their homeland and the cold lack of empathy from an international community that gazes on the Saharawi refugees indifferently while plundering their country.

Of course, Boisha's employment of the ready-made heart metaphor also serves a nation-building function: the poem performs a patriotic love of homeland. The single "engrav[ing]" in the penultimate stanza is perhaps the romantic peak of this loving sentiment. It takes time to carve a lover's name in stone. And such petroglyphs can endure for thousands of years. This stanza evokes the poetic voice's desire to declare his eternal love by giving a physical and permanent shape to an intangible emotion. Elsewhere, the *gallāba* give way to alluring imagery of a woman's chest, belly, and eyelashes, while the night is alive with beings "—rubbing [their] skin—." The use of a pair of em dashes around this phrase in the original Spanish version disrupts the stanza and thereby draws special attention to the parenthetical content. The playful eroticism undoes the supposed blandness of colonial-imagined deserts, suggesting desire and pleasure.

Boisha's personification of Western Sahara as a woman reinforces several of the poet's aforementioned claims to the territory, as well as his challenges to the discursive colonization of desert and wind. A cursory look at the literature on gender and nationalism shows the process of gendering the Western Saharan nation is nothing new.[20] However, the context of this poem, set against colonial terra nullius narratives that depict Western Sahara as an oversized, useless sandpit, gives a new meaning to a love poem where one's country is the sweetheart. The very idea of terra nullius is gendered: it implies a virgin territory waiting to be taken or impregnated, or rather developed and civilized, by colonial powers.[21] But this woman, this desert, is no virgin: the Saharawi poetic voice is already her lover. He knows and adores every part of her body—every corner of the Sahara—intimately. Miyek, for example, is compared to "a blemish / on the tummy of the earth." To know every blemish on the body of a lover is surely the superlative of intimate love. Furthermore, as is evident in the initial "A traveller asks me," the Saharawi poetic voice is explaining his relationship with his loved one, with Western Sahara, to an external interlocutor, perhaps even a would-be colonizer. He must therefore break down the negative stereotypes of desert that exist in the Western imaginary. Boisha does so by marking the various windblown land(marks) of Western Sahara—Miyek, Ziza, Tiris—and making them active and dynamic. The stone of the desert is transient. Craters emerge. The dunes touch. The physical, inanimate characteristics of the desert become self-moving, active subjects. By personifying the desert, Boisha first makes the Western reader see the life and characters of the *bādīa*, which are normally invisible for such audiences. As discussed, this is due to the Western imaginary's understanding of desert as a desolate, uninhabited, and bleak place (or non-place), or "something left to waste," to translate the Latin *desertum* from which the English *desert* is derived. Second, by giving agency to the desert's nonliving elements, Boisha simply and effectively hints at wider Saharawi epistemologies and cosmologies, which necessarily (in order for the continuation of Saharawi nomadism) respect other animals, organisms, elements, and minerals. He thereby highlights how the desert ecology works: humans, other beings, and elements of the desert atmosphere, such as wind and sand, live in a mutually beneficial, symbiotic relationship. One could draw a parallel here with Zapotec cosmologies, as explored by Howe. As she highlights, the Zapotec word for wind serves as a prefix for all nouns denoting living beings, and, relatedly, wind should be loved, cared for, and protected.[22] The implicit alternative to both Zapotec and Saharawi solidarity between humans and more-than-humans is the maintenance of the human–nature binary, the linked devaluation of nature, and today's resultant planetary crisis. Furthermore, using the windblown *gallāba* as the gravitational center of the poem serves to ridicule the corporate understanding of desert wind as that which is useless and burdensome until it is harnessed by man-made technology.

In her poem "The City of Wind," Fatma Galia Mohammed Salem likewise emphasizes, for a presumed Western audience (I make this presumption due to the poet's large serving of metaphors and similes from the European fairy-tale genre—for example, "The sun, my fairy godmother"), the interconnectivity and mutual reliance that exists between beings and things of the desert.[23] She, like Boisha, puts a question mark beside the anthropocentric ontology of petrocultural modernity. Mohammed Salem does this by making the desert not a passive intransient thing but rather a living subject that actively looks out for and elevates the emotional well-being of the poetic voice. Throughout the poem, the poetic voice is the object of transitive verbs for which desert elements serve as the subject: the stars watch over her, the moon spoils her, the sun guides and protects her, the desert makes her feel like a princess. The wind, however, is not the subject of a transitive verb but is rather her partner. It speaks with her, joining its voice with hers, in solidarity. Wind is her comrade-in-arms—her partner in the resistance. By making wind her (activist) equal, Mohammed Salem points to its role in the aforementioned desert ecology suggested at by Boisha. The picture she paints of a harmonious relationship between Saharawis, wind, and all the other desert elements serves to underline Siemens' contrasting alienation from the delicate desert ecology. While this poem is not written as a direct response to Siemens's propaganda on its developments in occupied Western Sahara, Mohammed Salem is aware of Siemens's activities there and shares news on campaigns against the company with her social media connections. She hints at an understanding and vocabulary of wind that contrasts those of Siemens by opposing the idea that wind should be dominated or tamed at all. As we have seen, drawing on the hegemonic colonial and capitalist understanding of energy, and indeed on terra nullius doctrine, Siemens depicts desert wind as underutilized: wind is the howling, gothic voice of the barbaric until corporate renewable developers tame it for electricity. Mohammed Salem undermines the colonial depiction of unharnessed wind by making apparent the latter's creative, productive role and power in the desert ecology, and thus in the lives and culture of Saharawi nomads. Wind, the poet shows, is an integral part of Saharawis' Western Sahara. If, as the UN suggests, the fate of Western Sahara should be decided by Saharawis,[24] and, as several court cases and a UN Legal Opinion suggest, the phosphates, fish, agricultural produce, and oil of Western Sahara can only be sold with the Saharawis' consent, then Mohammed Salem's foregrounding of the wind as her partner and equal suggests the same is true for "renewable" energy.[25]

> In the city of wind,
> I see what no one sees,
> I feel what no one feels,
> I say it, I repeat it with
> the wind and without regret
>
> The desert makes me feel
> like a princess
> in the city of wind.
>
> In the city of wind,
> there are palaces of stone
> and castles made of sand
> just like in the fairytales.
>
> The desert makes me feel
> like a princess
> in the city of wind.
>
> I walk barefoot on a rug of
> sand,
> smooth as silk
> and yellow as gold.
>
> I live under a great and vast sky
> veiled in blue,
> blue as sea.

In the city of wind
freedom was born without an owner.
When one lays one's eyes
on this distant and limitless horizon
the sight sails freely,
as if in a dream.

The desert makes me feel
like a princess
in my canvas palace,
surrounded by mirages,
like waterfalls and fountains,
that glide down from the mountains . . .
sparkling mirage, twinkling,
like the diamond that
suddenly springs up
from the depths
of the earth . . .

Earth, deserted and loved,
mother of the fire,
of the air, of the cold, of the silence,
of the nomad and of the wind.

The desert makes me feel
like a princess
in the city of wind.

The sun, my fairy godmother,
loves me, guides me
and protects me every day.

The moon, my magic mirror,
listens to me,
looks out for me and indulges me.

The stars, neighbours' lights
and towns of princesses,

nearby and far,
brighten for me every night,
they guard me and protect me.

The desert makes me feel
like a princess
in the city of wind.

In the city of wind
I see what no one sees,
I feel what no one feels.

I say it, I repeat it with
the wind and
without regret.

The desert makes me feel
like a princess
in the city of wind.[26]

En la ciudad del viento,
veo lo que nadie ve,
siento lo que nadie siente,
lo digo, lo repito con
el viento y no me arrepiento

El desierto me hace sentir
como una princesa
en la ciudad del viento.

En la ciudad del viento,
hay palacios de piedras
y castillos de arena,
como en los cuentos de hadas.

El desierto me hace sentir
como una princesa
en la ciudad del viento.

Ando descalza sobre una alfombra de arena,
suave como la seda
y dorada como el oro.

Vivo bajo un cielo
grande e inmenso,
cubierto por un velo azul,
azul como el mar.

En la ciudad del viento
la libertad nació sin dueño.
En este horizonte lejano y sin límite,
como un sueño,
cuando posa la mirada,
la vista navega libremente.

El desierto me hace sentir
como una princesa
en mi palacio de lona,
rodeado de espejismos,
como cascadas y fuentes de agua,
que se deslizan de las montañas . . .
espejismo brillante, con destellos,
como el diamante que
brota de repente
desde el fondo
de la tierra . . .

Tierra, yerma y querida,
madre del fuego,
del aire, del frío, del silencio,
del nómada y del viento.

El desierto me hace sentir
como una princesa
en la ciudad del viento.

El sol, mi hada madrina,
cada día me quiere,
me guía y me protege.

La luna, mi espejo mágico,
que me escucha,
me mira y me mima.

Las estrellas, luces de vecinos
y pueblos de princesas,
cercanas y lejanas,
cada noche me iluminan,
me vigilan y me amparan.

El desierto me hace sentir
como una princesa
en la ciudad del viento

En la ciudad del viento
veo lo que nadie ve,
siento lo que nadie siente.

Lo digo, lo repito
con el viento y
no me arrepiento.

El desierto me hace sentir
como una princesa
en la ciudad del viento.

Aeolian aesthetics are at work beyond just imagery and metaphor in Mohammed Salem's poem. She mimics the repetitive, cyclical whirlpools of wind in the structure and rhetorical strategies of the poem. Repetition is arguably the heftiest device in Mohammed Salem's poetic toolbox. She uses it to create stability, continuity, and a sense of control in knowing what is to come, thereby rupturing the disorientated, helpless, unpredictable, and chaotic trauma of war, occupation, and exile. The repetition of the longed-for "city of wind" and "desert," from which Saharawi refugees are exiled, invokes those places, making them poetically present. And the repetition of entire, and various, phrases—as if they were circulating in the wind—gives a chant-like quality to the aural poem, increasing its spiritual energy and power to affect the listener. One should bear in mind the Saharawi oral cultural legacy here. A nomadic existence explains Saharawi reliance on (oral) poets as transmitters of all types of knowledge, as anthropologist and fellow Saharawi poet Bahia Awah explores in some detail.[27] We can reasonably assume that Mohammed Salem's work was meant to be heard as well as read.

The aural aspects of the poem are further heightened by Mohammed Salem's use of devices such as alliteration and assonance (the stanza on the moon is a fitting example: "the moon, my magic mirror, / listens to me, / looks out for me and indulges me"). These are not just for (aeolian) aesthetic effect but also for reflecting the particular way that kinetic and sound energy work in the desert. There are relatively fewer obstacles to absorb the vibrations that produce sound waves in Western Sahara; thus, sound travels much farther and is more perceptible than elsewhere on earth. Some days in the *bādīa*, depending on the wind's direction and strength, one might be able to hear, for example, the rustle of marching ants on tree bark or the pitter-patter and hum of singing sand, blown and falling back on itself. Mohammed Salem's meticulous attention to the aural quality of her work serves to remind us, first, of Saharawis' aforementioned intimate knowledge of desert ecology and, second, of the essential role of wind for Saharawis' survival. For nomads, the windborne sounds are an aural guide—sentinels—to the life of the desertscape and can warn of nearing dangers from several miles away.[28] Nods to wind's enabling role in Saharawi nomadism challenges Siemens claim to "Ingenuity" in harnessing an otherwise "useless," "barbaric" wind of a desert rendered nullius.

As mentioned above, we cannot see the wind but only what is windblown and windborne. Aeolian aesthetics therefore appeal not just to vision but to all the senses through which we know the wind. I have discussed the rich aural texture of Mohammed Salem's work. Other than hearing its sighs, trickles, whistles, and roars, the main way we know the wind is through its touch.[29] In Mohammed Salem's case, her preferred tool for appealing to the tactile is synesthesia. In her stanza dedicated to sand, for example, she conjures the feel of its luxurious touch, as well as its rich sight: "I walk barefoot on a rug of / sand, /smooth as silk / and yellow as gold." She thereby communicates to a foreign audience that is presumably ignorant of the desert's delights the beauty and wonder of Western Sahara.[30] Much like Boisha's own evocation of touch then, Mohammed Salem seeks to challenge colonial mindsets that imagine the desert as bland and barren.

Again in the fashion of Boisha, Mohammed Salem conjures a particular windblown desertscape in order to persuade her audience of Saharawis' vast knowledge of, and place in, the desert ecology. With its golden and silky carpets of sand and sealike veils, Mohammed Salem's "The City of Wind" evokes an erg. An erg, or sand sea, is a large area of aeolian—that is, windblown—sand. The dunes of an erg are active, moving, and migratory. Ripples on their surface reveal the wind's direction and speed. The profile of dunes, whether they are curved into croissants, shaped like stars, or linear in form, tell nomads of the surrounding wind regime—its strength, pattern, and direction. Aeolian dunes are constituted by the wind, but they are also a mirror to it. They are weathervanes or wind roses.[31] As discussed above, nomads' ability to read the wind—whether live, as the wind blows, or by reading a past wind's tracks in the sand—makes it possible for them to use sound, scent, and the windswept as lookouts: they can tell what is coming and from where. Nomads also rely on reading the patterns made by the wind in sand to determine cardinal direction.[32] Using wind roses etched in their minds, Saharawi nomads can read the wind, or its tracks in the sand, like maps. The spine of a dune might point to Mecca at sunset prayer time. The feel and warmth of a moving air mass might signal a route to a well.

The only erg or sand sea in Western Sahara sits in the country's heel, within the wider beloved Tiris region. The erg, known by the name of its elevated *galb* Azefal, runs from northeast Mauritania, through southwest Western Sahara, and back into Mauritania again. Here is Mohammed Salem's oxymoronic city of wind. She gives the famed Saharawi erg the label of "city" to show her lack of want for colonial-imposed urbanism, modernism, and so-called development. The rural city of wind offers all she could need. Likewise, the poetic voice uses all the senses to show us the desert is as valuable as "silk," "gold," or "diamonds" to Saharawis. Her palace is her animal-skin tent, and the desert—not Siemens's contribution to so-called economic development—provides all the riches she desires. As the author told me in an interview, "We don't need big companies. All nomads need are oases, water, vegetation, and the natural wind."[33]

Mohammed Salem's rejection of economic development and urbanization—made clear in her celebration of a humble, nomadic desert life free of material wealth—can be read as a wider refutation, or poetic silencing, of petroculture. To draw on Dorothy Odartey-Wellington's work on the near absence of the Moroccan military wall from Friendship Generation poetry, a *purposeful lack* can be read as resistance. Odartey-Wellington argues that the Saharawi "writers' imaginations make for" the Saharawi heartlands, and indeed its beaches, "without any reference to the wall between the poetic voice and the ocean."[34] Odartey-Wellington suggests this is because the writers "are engaged in a poetic negation, or undoing, of the power of the wall."[35] The poets thereby subtly remind us that the walls of occupied Western Sahara, both the military ones and the fences surrounding the king's and Siemens's vast wind farms, are not, after all, the "benevolent enclosures" that the terra nullius narrative promised. Likewise, the purposeful lack of petrocommodities and the "neoliberal freedoms" of petroculture in Boisha and Mohammed Salem's work, as well as their reliance on aeolian, rather than oil, aesthetics constitute a (temporary) barrier to the pernicious reach of petroculture.

Conclusion

Wind energy developments in Western Sahara are a case study in energy colonialism. Saharawi poetry, with its incorporation of aeolian aesthetics, shows a way to challenge energy colonialism on an artistic level. Saharawi Friendship Generation poets create melancholic pieces that long for a return to a pre-petrocultural, nomadic way of life. Saharawi culture is entangled with the wind, and the homeland that Saharawis love is materially shaped by it. Emerging from this culture are what I call *aeolian aesthetics*: poetry and visual art informed by wind in their structures, motifs, imagery, and rhetorical devices. Aeolian aesthetics are characterized by a decided appeal to the senses through which we know the wind: sound and touch, and visions of the windblown. On a political level, they undermine hegemonic understandings of energy, which make nature (in this case, wind) something to be harnessed and dominated in order to power capitalism and colonialism. The poets do so, first, by showing wind's role as a creative force, both in the artistic sense and in a geological sense. The three desertscapes invoked in the poetic texts explored here—the hated plains of the Algerian Hamada, the cartographic possibilities of the *bādīa*'s *gallāba*, and the dunes of the erg surrounding *galb* Azefal—are products of ongoing aeolian processes. Second, the poets show the wind's productive power by nodding to how it has historically shaped, and continues to enable, the Saharawi's nomadic existence. Third, they claim the desert's winds as an integral part of Western Sahara, and therefore as a subject of Saharawi sovereignty, for purposes of wind's exploitation as energy. Aeolian aesthetics then, in the particular case of Western Sahara, are employed in an anticolonial movement at the poetic level. Yet the ability, demonstrated by Boisha and Mohammed Salem, of aeolian aesthetics to undermine the logics of petroculture itself indicates their wider possibilities for those living in petroculture and wishing to resist it. In response to the current environmental crisis, Imre Szeman and Dominic Boyer argue that the task of the energy humanities is to "grasp the full intricacies of our imbrication with energy systems (with fossil fuels in particular), and second, map out other ways of being, behaving and belonging in relation to both old and new forms of energy."[36] Aeolian aesthetics allow us to envisage a social existence—forged around a potentially renewable energy source—that refuses the colonial, capitalist clothes handed down from petroculture. A non-colonial, or anti-colonial, wind imaginary can shape a fairer energy system.

CHAPTER 6

The *Īrīfi*

Wind as Harbinger in Arabic-Language Cultural Production

"Khadija always seems to have watery eyes, as if she has just been crying."[1] When Abdessamad Al Montassir asks Khadija about what happened in 1975, she finds herself unable to speak. In the short documentary *Galb'Echaouf*, we do not see Khadija's face—just her orange, hennaed fingernails and the folds of her *melḥfa* over her legs. These images, as well as the *Ḥassānīya* words of her speech, tell us that she is Saharawi. Khadija suggests to Al Montassir, "Go and ask the ruins, the desert, its thorny plants. They saw and lived through everything, and have remained there. They can better tell of what happened."

Abdessamad Al Montassir is a multimedia artist who was born in Boujdour in 1989. He follows Khadija's advice, not just in the film in which she appears, but in several of his works. Al Montassir converses with desert nonhumans as well as with other Saharawis who, like Khadija, are old enough to not be able to speak of 1975. In *Galb'Echaouf* and the multimedia installation *Al Amakine*, the Saharan winds fill in the silences of humans.[2] They speak of horror.

In this chapter, I further explore an Indigenous, anti-colonial wind imaginary that provides an alternative to the colonial wind imaginaries of eighteenth- to nineteenth-century mariners, Spanish colonizers, and modern-day energy multinationals. I do so here by moving from a focus on Spanish-language Saharawi poetry to Saharawi imaginations of the wind in Arabic (*Ḥassānīya* and Darija dialects) language cultural production. I first look at Al Montassir's multimedia work and then move on to explore representations of wind in poetry, both of traditional nomadic poets Zaim Allal and El Badi and more modern iterations by Hamza Lakhal, Malainin Lakhal, and Ahmed Ould Mohammed Fadel. In Arabic language cultural production, I argue that Saharawis give wind agency, personality, and value. They do so in order to showcase Saharawi knowledge of wind, the aeolian and its relationship with desert ecosystems, and also to urge solidarity with the Saharawi cause.

What the Wind Knows

Al Montassir's short film *Galb'Echaouf* features interviews with three human subjects. Facing their inability to voice their memories, Al Montassir turns in the film to nonhumans of the desert—plants, ruins, rocks, the wind—in search of knowledge of what happened in 1975. One key interviewee is the daghmous (Euphorbia Echinus), a thorny plant that is autochthonous to Western Sahara. According to Saharawi legend, it once wore evergreen leaves and beautiful flowers instead of spines.[3] It developed thorns in response to threat and "withdrew into itself."[4] "No one knows what happened to it," says the narrator, with the plant's silence and implied experience of violence reflecting the story of Saharawis.[5]

We, the viewers, look at boulders, fossils, and engraved words on the rocks of ruined buildings—these too are nonhumans that speak of the past. As we hear Khadija's voice recalling her childhood as a nomad and the journeys she made, close-ups of her fingers flexing as she talks are juxtaposed with lingering shots of such rocks. The effect is to make the rocks speak with Khadija's voice. They tell of when nomads and rocks dwelt together in the desert before the occurrences of 1975.

It is the wind that replaces Dah's unsaid words. Two sequences feature this elderly man, sitting alone in a *ḫaīma* in the desert. In the first, he tells us that he is not going to say anything—that he "can't talk about it anymore." The camera peaks over his shoulder, peering out to the desert, as he speaks. We hear a gentle, swooshing wind and birdsong—the soundscape of a calm desert. Dah stops speaking. The wind blows a little more forcefully now, causing the quickening flap-flap of the *ḫaīma* entrance's edges. It is as if the wind has stepped into the *ḫaīma* to be at Dah's side.

In the second sequence featuring Dah, the wind is with him.[6] Its sound becomes overbearing. Its crashes increase in intensity and volume, as if rising up the Beaufort scale, as the sequence progresses. In the first shot, Dah's head hangs forward toward the floor. A series of subsequent close-ups communicate unspoken trauma. He rocks back and forth. A fly crosses his face and meets his eye. He does not bother to brush it away. The fingers of his clasped hands are tense. They look arthritic, wooden, and hurting. The camera cuts to Dah's face.

The inner edges of his eyebrows point upward in an expression of sadness. His eyes move to look at something. We follow his gaze out to the desert. By now, the wind sounds like walls of waves heaved onto rocks, as if the listener/viewer is caught in a storm at sea. Then, on top of the crashing, comes an unsettling—no, a terrifying—low moan, a slow *wow*, *wow*, like an air raid siren. Now the camera cuts to a wide shot of the desert. Ribbons of twisting sand glide eerily toward us as yet another layer of sound—a ghostly, high-pitched wind chime—completes the spine-chilling chorus. This is poor Dah's testimony, told with dramatic effect by the wind.

Al Montassir's use of the voiceover technique to make rocks and plants speak in the timbres of Khadija, and soundtrack juxtaposed with Dah's image to make wind speak the old man's silent sufferings, suggest an entanglement of Saharawis and more-than-human elements of the desert. The final interviewee, a young man called Weld Sidi, offers perspectives that further blur the boundaries between humans and their others. The narrator, speaking of Weld Sidi, says, "He used to tell me that our bodies were like plants, which lost their leaves but remained standing, that our sparkling eyes were like stars, which though dead, still twinkled."[7] Weld Sidi himself tells us, "The plants and the mountains remember our traces, and our stories spread into places we have not yet travelled through."[8] After sharing Weld Sidi's wisdom, the film closes with still images of plants, trees, and thorns in night's darkness, fading to black, with wind swooshing all the while and on through the credits. This ending ensures an enduring unsettledness. We do not know, from the film, precisely what happened to Khadija, Dah, and others of their generation in 1975, but we are sure that it was horrifying.

My ear, though unattuned, can notice the differing voices and moods of the winds in the film. The use of winds' moods reminds me of Saharawi nomads' far more acute (learned) sensitivity to, and knowledge of, the winds and how such knowledge makes their life in the *bādia* possible. But the particular role that Al Montassir gives to the wind also suggests a different way to understand nomads' knowledge. The film proposes that the wind *knows*. It suggests that Saharawi nomads not only know the wind, but they also know how to *listen* to the wind—to its advice, warnings, and predictions. The wind is intelligent. Al Montassir says of his work, "Most of the time when we try to think about nature, we think about it in human terms. But what is crucial for me is to explore the possible emergence and surging of potentials that we cannot anticipate at the human level."[9] Al Montassir not only underlines the mutual dependencies and interrelations between Saharawis and desert more-than-humans, but he also highlights the intelligence of the latter. If we think of the wind as more-than-human, and of adjectives that personify the wind or verbs that imply its agency as more than poetic devices, we become more aware of humans' entwinement with the wind. Maurice Merleau-Ponty's work on the phenomenology of perception is useful here.[10] According to Merleau-Ponty, humans are possessed by what they perceive. Perception involves a mutual interaction between the perceiver and the perceived; therefore, both are agents. If you feel the wind on your body—its push, its temperature—the wind is providing you with those sensations. It allows you to feel, and you are in a reciprocal encounter with it.

Letting wind, rocks, and plants tell of the horrifying injustices that authorities inflicted on the Saharawi population is an ingenious challenge to Morocco's silencing of the situation in Western Sahara. The film avoids the naming, in human languages, of Western Sahara, and, of course, the human silence on political violence is the bedrock of the film. The human silence signals not only trauma but also arguably nods at the completely taboo nature of this history (and present) in Morocco and Moroccan-occupied Western Sahara. Al Montassir has described his intention with the film as to "tell the story without telling the story" and to "make a stand without taking a stance."[11] This allows him to avoid not only the censors but also the potential alienation of an audience that may initially be antipathetic to Saharawi perspectives on the conflict.

A key target audience for Al Montassir is the Moroccan public, which is reflected in his choice of dialect for the narration of the film. The narrator of *Galb'Echaouf* is imagined to be a visitor to the *bādīa,* guided by his Saharawi friend Weld Sidi. Al Montassir provides the narrator's voice himself and speaks in Moroccan Darija. In doing so, he represents his Moroccan audience, often unaware of what happened to the Saharawis immediately following the Moroccan invasion of the Sahara, and what he wishes them to do after viewing the film: "go and research what happened and find their own answers."[12]

European colonialists have long implied that Indigenous populations are "closer to nature" and are therefore uncivilized, uneducated, and barbaric.[13] By blurring the boundaries between Saharawi humans and other desert dwellers, Al Montassir does not make Saharawis "closer to nature" according to the colonial world of binaries but rather challenges these binaries and raises the status of all the Sahara's constituents. The wind, rocks, and trees experience trauma, remember, and communicate. Desert more-than-humans are also shown as unique sources of knowledge. They have value beyond the square kilometers available for solar farms and beyond the kilowatt hours of energy produced from the movement of air.

The stars of *Galb'Echaouf* are also the protagonists of Al Montassir's multimedia installation *Al Amakine* (Arabic for "the places"). The installation is composed of photographs of Western Sahara's *bādīa* and portraits of its plants, displayed on lightboxes. A sound installation accompanies the photographs. Al Montassir describes his subjects as "non-material archives," while he wishes the installation as a whole to be "an opening towards stories and geographical spaces outside official maps and histories"[14] and "an endemic cartography of traumatic events" in Western Sahara.[15] Al Montassir's work involves a collaboration with biologists, who measure the plant's acidity. This makes it possible to date the stress and trauma the plant has experienced—the daghmous offers knowledge on past events where humans cannot. The narrator underlines this in his voiceover: the desert plants *know*; it is just that they "speak in a temporality that eludes us."[16]

The sound installation samples several Saharawi poems that, in Al Montassir's words, "take into consideration the non-human elements (plants)."[17] One poem is by Zaim Allal, which, for ease, I will call "Acacia Leaves." In "Acacia Leaves," the wind represents all the challenges—political, bellicose, meteorological, and others—while the Acacia tree is a metaphor for Saharawis and their national project. Its fallen leaves, blown in the wind, stand in for Saharawi "traitors."[18]

> Acacia leaves fade, turn yellow,
> Fall and blow in the wind.
> Then new leaves turn green and grow,
> Yet, like the first ones, they wither and fall.
> Only the long-suffering Acacia tree remains
> Undefeated and steadfast against the wind.
>
> The strong wind drives plant litter away.
> I wonder, how can leaf litter stay?!
> That is how most faint-hearted men
> Fall down one by one.
> And when the battle cry echoes through the violent winds of war,
> They will not be found.
>
> Acacia leaves fade, turn yellow,
> Fall and blow in the wind.
> Then new leaves turn green and grow,
> Yet, like the first ones, they wither and fall.
> Only the long-suffering Acacia tree remains
> Undefeated and steadfast against the wind.

The weak crumble
And go down without a fight.
Like drought-starved sheep
Or an unstaked tent.
And when the war rages on,
And its drums are beating,
You will witness how their honour easily falls
And gets trampled, like dead leaves, on the ground.
A dried-up leaf,
Does not need a strong wind
To wrestle it from the tree

Acacia leaves fade, turn yellow,
Fall and blow in the wind.
Then new leaves turn green and grow,
Yet, like the first ones, they wither and fall.
Only the long-suffering Acacia tree remains
Undefeated and steadfast against the wind.

And there is the social climber who tries to get rich
Through trading in his people's destiny.
Who ignores the will of the martyrs,
The missing, and the war-wounded.
A people still fighting lawlessness,
Bearing the flag of rights and bright justice.
The traitor who strikes deals with the invader
Walks into the wind at night.
He wallows in a swamp of dishonour,
And accepts the Mendrish's deal
Hoping to be an important man.
But the saying goes:
An honest man does not make himself a dog
For the sake of a bone
This social climber, like leaves, after all,
Is doomed to fall.

Acacia leaves fade, turn yellow,
Fall and blow in the wind.

Then new leaves turn green and grow,
Yet, like the first ones, they wither and fall.
Only the long-suffering Acacia tree remains
Undefeated and steadfast against the wind.[19]

يلوا ورڨ الطلح ويصفار *** يطيح وتوغد بيه الريح
وينبت ورڨ اجديد ويخظار *** ويلوا كيف اللول ويطيح
ويتم الطـــــلح الا صبار*** فعناد الريح اوڨوف اصحيح

*****

يصوڨ أزفزاف الحشلاف *** وأسفساف افوجه ازفزاف
ما يوڨف والرجلي لخفاف *** الا لكثر فيه الطاييح
ولكثر فيه الا ما ينشاف *** تو الريح وساعة لبريح

*****

يلوا ورڨ الطلح ويصفار *** يطيح وتوغد بيه الريح
وينبت ورڨ اجديد ويخظار *** ويلوا كيف اللول ويطيح
ويتم الطـــــلح الا صبار*** فعناد الريح اوڨوف اصحيح

*****

الناس اضعيفها ينساس *** ويطيح امسيكين بلا ساس
كيف اواش اسوامو ترخاس *** فالشدة وزنو ريح اسنيح
اتراه إلى حركو لقراس *** اشبيح اطبول الشر الشيح
ما يبني واللي ماهو عاس *** على عرضو، عرضو فطريح
وذاك النوع الناس يباس *** ورڨو ماهو محتاج الريح

*****

يلوا ورڨ الطلح ويصفار *** يطيح وتوغد بيه الريح
وينبت ورڨ اجديد ويخظار *** ويلوا كيف اللول ويطيح
ويتم الطـــــلح الا صبار*** فعناد الريح اوڨوف اصحيح

*****

بالطوطيح الڨام يهايت *** للرزق ايتيفي بعنايت
شعبو وامفرط فوصايت *** شهيد وفقيد وجريح
شعب ايكافح رافد رايت *** حق وعدل امبروك صريح
حد اتعاكب عنو بايت *** فالريح وبايت فاتمرميح
لفظاحة قانع بكرايت *** بخنوس الناس يدور اشبيح
جلدو ميت ڨالو فايت *** عن ما زين على الكلب انبيح
ماهو فمراحو واروايت *** رزق يجي بالطوطيح اشحيح

*****

يلوا ورڨ الطلح ويصفار *** يطيح وتوغد بيه الريح
وينبت ورڨ اجديد ويخظار *** ويلوا كيف اللول ويطيح
ويتم الطـــــلح الا صبار*** فعناد الريح اوڨوف اصحيح

Here, the wind is an agent that sorts the brave from the cowardly and the treacherous from the loyal. Allal uses Saharan winds to explore themes of weakness and betrayal. In doing so, he alludes to Saharawis that, tempted by Moroccan state's bribes, leave behind "the cause." The small numbers of Saharawis that accept payments and/or positions of power in return for shifting their allegiance to Morocco are known as *'a'idin*. The *'a'idin* are generally offered subsidized housing and a monthly stipend in exchange for renouncing their support for Saharawi independence.[20] Higher-profile Saharawis are offered significant political positions and grand economic incentives. The most famous of pro-Moroccan Saharawis are those at the center of the *makhzen* (Morocco's ruling class), such as the Derhem family. Atlas Sahara, now owned by Derhem Holdings, today has an almost total monopoly over petrol stations in occupied Western Sahara. According to Western Sahara Resource Watch, Atlas Sahara controls forty-seven petrol stations in the territory and several depots. The Derhem brothers are also active in the fisheries and agricultural industries.[21] In 2006, Mohammed VI appointed Hassan Derhem, one of the richest men in occupied Western Sahara, to the Royal Consultative Council for Saharan Affairs (CORCAS). Rkia Derham was the Moroccan Secretary of State for Foreign Trade until 2019. In 2014, Western Sahara Resource Watch estimated the amount of petroleum delivered annually to occupied Western Sahara at between 250 and 280 million liters.[22] As I mentioned in the introduction to this book, in 2020, most petroleum products used in occupied Western Sahara were exported from Teesside and Immingham, UK. Once brought on land at the port of El Aaiún or Dakhla, there are no pipelines to transport petroleum; thus, it is stored in depots. The Moroccan army, for example, has numerous modest fuel storage facilities along the length of the wall. Morocco imports diesel fuel for use in electrical generation stations, liquid propane gas (LPG) for heating and cooking appliances, petrol for motor vehicles, and small amounts of other specialist petroleum products for use in machine lubricants and aviation fuels. Wind imaginaries, the energy system, politics, and cronyism are all tangled up in occupied Western Sahara.

I ask Hamza to help me translate Allal's poem, most of which, with its vocabulary so precise to Saharawi nomadic culture, I barely understand. Hamza takes two weeks to complete the first draft. He too struggles with the vocabulary. He visits elderly relatives—former nomads—who can help. He also calls Zaim Allal, who can explain the intended meaning of the most ambiguous words (Allal also gives Hamza an extra verse to add to the poem—Saharawi oral poems are like living beings: growing and changing in form and color with each performance). Sam Berkson and Mohammed Sulaiman, other translators of Saharawi poetry from *Ḥassānīya* to English, have encountered similar challenges. They joke of "Rabouni *Ḥassānīya*"—that is, the version of *Ḥassānīya* spoken in the refugee camps (hence the use of the name of the camps' capital Rabouni) versus the *Ḥassānīya* of nomads. The latter, says Berkson, is gradually losing much of its nomad-specific vocabulary due to decades of forced settlement.[23] In the translation of "Acacia Leaves" above, we lose specific named winds such as the *Azafzaf*, which nomads know to uproot trees, and the *āwāsh*, which means both "the dregs of society" and a "herd of sheep weakened by drought," but we keep the Mendrish, a famous squirrel from Saharawi folktales that is tricked by greed. We also gain a Danish proverb (an honest man does not make himself a dog for the sake of a bone).

Some say it is impossible to translate a poem. Rather, the translator creates an altogether new poem, inspired by the original one. Syntax, meaning, rhythm, metaphor, aesthetics, or other poetic aspects may be sacrificed in the translation or recreation, but others emerge in the new version. There is an added complication to this loss (and gain) when it comes to translating oral poetry from *Ḥassānīya*. I mentioned above that I call Allal's poem "Acacia Leaves" for convenience. *Ḥassānīya* poems do not usually have titles. Listeners, not poets, assign poems temporary titles according to what the poem means for them personally—depending on the reader, poems, like the winds, can change direction. Listeners might pick the first line of a poem and use it as a title, or they might use whichever theme in the poem speaks most strongly to their soul. The poet may add verses, or change their wording, each time they speak the poem. By transcribing Zaim Allal's poem and the title that Hamza has chosen, we freeze its previously transitory, flexible, or *nomadic* meaning and form. However, the translated poem offers a window into Saharawi culture and the Saharawi cause for outsiders. It becomes a messenger.

Solidarity and the Second Person

A common use of second person pronouns in several languages is to establish solidarity between the speaker and their interlocutor(s).[24] In Arabic, there exists the *iltifāt*, a feature especially common in early Arab poetry and in the Quran, in which the grammatical person changes abruptly. Geert Jan van Gelder describes the purpose of the *iltifāt* as follows: "The shift of grammatical person results in a shift of focus that may unsettle the audience, which is invited to feel itself in the role of the observer or the addressee, even to identify with the 'poet' and to share his experience."[25] Modern Saharawi poets composing in *Ḥassānīya* use the second person and the *iltifāt* and strengthen their effects by using wind to the same end.[26] The idea of wind as messenger is common to many cultures. The four winds of both English and Spanish idioms of dispersal (to the four winds/*a los cuatro vientos*) appear in the bible, in which wind—and specifically desert wind—features as a messenger as well as a source for allegorical imagery. In Islamic traditions too, winds and clouds can be signs from God. In his first collection, *A Destiny in Windblown Poetry*, Lakhal conjures his verses from the wind.[27]

Here are a few facts about Hamza: He was born in El Aaiún city. He is a poet. His greatest influence is Palestinian writer Mahmoud Darwish. His birth date in his passport is different from his actual birthday. Among his favorite songs is Fairuz's "Neighbour of the Valley." "It makes me think of the valley on the opposite bank of the Saguia el Hamra river, as it passes through El Aaiún." *A Destiny in Windblown Poetry* is banned in Morocco and occupied Western Sahara. I therefore receive my copy—ordered from its Paris publisher, Harmattan (named after a Saharan wind, curiously enough)—before Hamza ever holds the book in his hands.

Wind is at the center of the collection. The second half of its original Modern Standard Arabic title is a play on words around the Arabic idiom *fī mahab ar-rīḥ* ("into the wind"), which, rather like the English idiom (although without the connotation of being caught in an argument specifically) "into the crossfire," denotes danger. The idiom has a second meaning: it describes the subject as abandoned and unremembered, as if blown away in the wind. With this title, then, Hamza brings to the fore the risky and precarious situation of Saharawi poets living in occupied Western Sahara, as well as the forgotten nature of Western Sahara's cause in the Global North.[28]

Here follows an extract from Lakhal's poem "Songs of Exile":[29]

You will remain alone
Lonely as you are
From the wind you come
To the wind you walk
A stranger, like a homeless grain of sand
In the wind you grow
In the wind you suffer
A fugitive, like a rhyme in a poem.

You ask for the land of Saguia
You ask the clouds for your beloved land

"Is my homeland's dawn ahead?
Answer me . . . answer to revive me
Answer me . . . answer to cure me
My country is my wound
My wound is my country"
But no one responds.

The clouds stayed silent
The dawn did not come
Questions brought no healing
No land for you.
The wind is your home
Your exile
Your shelter
Your life
Your stream
Your path

وحيداً ستبقى
كما أنت وحدكْ..!

من الريحِ تأتي
إلى الريحِ تمشي
غريبا كحبَّةِ رملٍ شريدهْ
وفي الريحِ تكبرُ
بالريحِ تشقى
طريداً كقافيةٍ في القصيدهْ
وتسألُ عن أرضِ ساقيةٍ
تسألُ الغيمَ عن أرضِ عشقكَ
—هل لاحَ صبحُ بلادي ؟
أجبني . . . أجبني لأحيا
أجبني . . . أجبني لأشفى
بلادي جراحي . . .
جراحي بلادي . . .
وما من مجيبْ
فلا الغيمُ باحَ
ولا الصبحُ لاحَ
ولا السُّؤْلُ داوى الجراحَ
ولاأرضَ لكْ

بيتكَ الريحُ..
مَـنْفَـاكَ.. /
مَــأوَاكَ.. /
مَحْـيَـاكَ.. /
مَجْــرَاكَ.. /
مَسْــرَاكَ.. /

The anguish of exile is physical pain in this poem. The wind becomes a medium in which the exiled second person dwells, building on the idiom referred to in the title of the collection, in which wind is linked to peril and to the forgotten. The first verse that I cite here ("You will remain alone / . . . / A fugitive, like a rhyme in a poem") is a refrain that is repeated several times throughout the poem—as if it is itself the "homeless grain of sand" swirling in the wind. More-than-human desert existents such as the clouds are agents with whom the poem's second person engages, nodding to Saharawi nomads' knowledge of the interconnected lives and activities of all the desert's elements, and the lack of hierarchy between them.

The exiled speaker of the poem asks the clouds about the future of his country. Here, Hamza nods to a famous self-nickname of the Saharawi nomads: "the children of the rainclouds (*āūlād al-mizna*)."[30] The nickname comes from nomads' ability to read the clouds to predict natural phenomena and their propensity to track clouds in pursuit of water. As Hamza himself adds, "Clouds move forward, they push the world's time on."[31] Wind in this poem represents the various international forces that "push" Saharawis from the international community to the occupation to the desire for revolution, and so on. The various winds clash and push the Saharawis anywhere but their homeland.[32] As the title of Hamza's first collection suggests, and as is the thesis of my book, the Saharawis' fate is tied up with the Saharan winds.

For Ahmed Ould Mohammed Fadel, the wind, specifically the *Gblīya*, carries messages. He wrote an ode to this wind—an extract of which is below—while in a hospital in Spain. Upon initiating treatment, he was told he would have to stay for six months. During the first month, feeling homesick, he wrote the poem, in which he addresses the *Gblīya* in the second person, and urges it for news from Western Sahara.[33]

O south wind,
Bring me news of my beloved southern land.
Bring it to me, please,
As fast as you can.

The south wind coming, these days,
Here in Spain
Makes me feel a kind of unease
I've never experienced before.
What kind of feeling is this?
Even the clouds suffer.
But the gentle breeze,
Which the south wind brought,
Showed me my beloved places
And occupied my thoughts.
Then, it took me on a journey
Through beautiful tracks,
Colorful small mountains, and valleys.
And it took me back to this place
Where fate placed me.[34]

يا الڭبلية الله يجازيك *** هذي لرض الڭبلية
تمي لريتي من لخبار *** عنها رديه اعليا

★★★★★

اطلوع الڭبلية ذو ليام *** اعليا ساحل فتراب اسبان
عاد امطريلي شي ما كان *** من كز الحزم اعليا
وعاد امطري للنو احزام *** من لرياح لماهي ڭبلية
ما طراتو يغير النسيم *** لطلعت بيه الڭبلية
ظهر لي لوكار إلين *** اڭطع لي تخمامي بيا
وسدرني فمسارب زينين *** وڭلابَ زرڭ وفاودية
ورجعني لبلد لمنين *** طرحتني فيه الكتبية

Mohammed Fadel's poem shows, once again, wind's common role as messenger in Saharawi poetry, reflecting nomads' ability to use the wind to foresee natural phenomena, threats, or good news. The *Gblīya*'s provocation of nostalgic feelings in the heart of the ill exiled man likewise calls for the reader's empathy and highlights the Saharawi community's *Gblīya*-induced emotional sensibilities. Writing or speaking to the *Gblīya* in the second person allows us, the readers or listeners, to identify with it as if it were a person. It urges solidarity.

Recalling the *Bādīa*

The poets are in the *bādīa*. It is spring in Tiris. The pastures are green and the wells are freshwater and full. Beautiful, beloved Tiris. If you pronounce it like a *Ḥassānīya* speaker, with the *i*'s pronounced like the *y* in *happy* and the *r* with an almost-trill, the corners of your mouth cannot help but curve into the hint of a smile. *Tiris*.

Moloud[35] and his wife Khadijatou[36] were born nomads. As a young man, Moloud got a job in the Indigenous unit of the colonial police force. He summarizes his memories of the Spanish colonial era in El Aaiún city with one word: *inequality*. Khadijatou lived nomadically in Tiris, Western Sahara's heartland in every sense of the word, until a teenager. They are the parents of my hostess, with whom they share a courtyard in El Aaiún camp, Algeria. Most nights, we sit there together with a tea stove as the sun sets, and then, after dark, we share dinner and talk until sleep comes. Moloud tells me about the Sahara's winds and how to use them to navigate at night. Khadijatou entertains us all with traditional tales of the desert's nonhuman animals (such as Mendrish, the greedy and gullible squirrel that features in the quoted Allal poem above) and their shenanigans. Other nights, I sit inside, where there is electric light, in order to read and work with Gajmola,[37] Moulud and Khadijatou's youngest daughter. The evening before our meeting with Zaim Allal, Gajmola is so excited that she is sure she will not sleep. He is, after all, one of the SADR's most celebrated poets. She brings in one of Allal's collections from her brother-in-law's small library in the next room. We take turns leafing through it, skimming the poetry, and pausing and lingering on poems that speak to us loudly.

We meet Zaim Allal at a Ministry of Culture festival. He wears a sky-blue *darrāʿa* and a silver beard and welcomes us in a cool office away from the growing hubbub outside. The wall has a mural: dozens of women carry Saharawi flags and wear brightly colored *āmlāḥaf*—jade and silver stripes, shocking pink roses. One has psychedelic orange swirls. Allal apologizes that he is short of time. He works for the ministry now, and today is a busy day for a civil service poet. "We use poetry to encourage people to resist these trials of nature, to stay here until we can go back to our land."[38] This was not always the purpose of Saharawi poetry, of course. Allal was born in Hausa in 1957 into a family of nomad poets. In those days, his relatives would celebrate the "beauty of nature" but also use their verses to offer advice on nature's "challenges" and "opportunities." He offers an old verse as an example. "It is a riff on a Saharawi proverb," Zaim tells me. The verse (and the proverb) recommends that "on the occasion of rains in summer, you should sell your harvest and invest in livestock."[39] Poetry is born of nomadism and simultaneously facilitates it.

Later in the day, Allal will recite some of his work in a carpeted *ḫaīma*, opened at one side to create a stage. The latter is fitted with quilted leather cushions dyed red, yellow, and green; sweet, splashing tea glasses on glowing charcoal braziers; and applauding hennaed hands belonging to audience members who are also performers. Some of the latter play the *ṭbal*—a traditional women's drum—and ululate. Tea, music, poetry, desert, and nomadism go together—each one shapes the other.

Ḥassānīya poetry is a stage for the aesthetic rewards of the desert. Allal claims that Saharawi nomads' movement across the desert heartlands is the source of their poetry.[40] As they move, nomads reflect, in verse, "the sounds made by the wind, the footsteps of camels over the sand, the rattling of leaves and the noise made by raindrops as they fall over caves."[41] Allal is an army man. He joined the armed revolution in 1975 and worked intermittently (he also studied in Algeria) as a soldier throughout the eighties.[42] While today, for Allal, poetry is "a weapon in the war with Morocco," the aim of poetry before the conflict emerged was to discuss "how to face the natural environment."[43] This included celebrating beautiful places and phenomena in the desert or reciting odes to helpful winds such as the *Gblīya*, and also transmitting knowledge concerning all aspects of life in nature or nomadic life, from the best treatments for each ailment suffered by camels to how to navigate to pasture-rich regions and when.

Around the world, Indigenous humans have preserved specific understandings, rooted in their cultures, of their local ecosystems, which guide relations between humans and more-than-humans.[44] Such understandings are commonly termed *Indigenous knowledge*. Nguia Hassan, the environmental activist and nomad whom I quoted in chapter 2, sees Saharawi ecological knowledge as tied up with Saharawi belonging: "Despite the fact that nomads have not been to universities and so on, they were smart enough to predict the rain. This is because they belong to this environment."[45] As Deborah Bird Rose points out, if we simplify the sustainable practices of some Indigenous peoples, we may risk reinforcing the colonial culture/nature binary in which Indigenous peoples are seen as "at one" with nature, and reinforcing the discourse of the "noble savage."[46] Yet, as Bird Rose also highlights, research into various Indigenous communities' relationships with the nonhuman world show that such communities have lived sustainably, have maintained their ecosystems in a state of balance, and have lived by philosophies that emphasize "long-term relationships of mutuality between humans and non-humans."[47] For Saharawis, the winds are a key part of their ecosystem. Their knowledge of this ecosystem is passed intergenerationally via poetry. As Abram points out in more general terms, oral cultures preserve practical knowledge in poetry, the rhythmic nature of which makes it easier for people to recall.[48] Or, in the eloquent words of Jane Belfrage, "It was in the soundscape that Indigenous people's knowledges were published."[49]

Saharawi Friendship Generation poet and anthropologist Bahia Awah takes poetry as his map in his book *Tiris: Literary Routes*. One might describe this beautiful book as an ethnography of Saharawi poetic cartography and topology, in which Awah recalls a journey through Western Sahara's free zone, visiting nomad poets and locations of poetic and national importance for the Saharawi imagined community. Awah implies that Saharawi geography and Saharawi poetic traditions are interlinked disciplines.[50] In another article coauthored with fellow anthropologist Juan Carlos Gimeno Martín, Awah highlights how "poetic games" in the works of poets such as Mohamed Moulud Budi (Beibuh), Badi Mohamed Salem, and Sidi Brahim Salama Echdud highlight Saharawi toponyms and the relations between them.[51] Rodriguez Esteban argues that nomads possess "mental maps" of the desert thanks to years of poetry recitation. His explanation is worth quoting in extenso:

> Currently, when you travel with Saharawi guides and you have maps that are constantly positioned using GPS and you ask them, often knowing the answer, what lies ahead in different directions, their answers are incredibly exact. Sandstorms can last for several hours or days. All visual references are hidden and orientation when setting off is always tricky. But the guides examine clues, such as the direction of roots of plants like the *fredolia aretioides*, and get back on track quickly. In their stories about happenings during the war with Morocco, they situate events in various desert settings that are different for Saharawis but that would be indistinguishable for a European. A possible conclusion is that poetic games are a good start in drawing the mental maps that everyone needs to orientate themselves: mental maps that are gradually filled during long migrations, in the solitude of the desert, without distractions, via memories and by listening to and reciting poetry.[52]

Boisha's poem "Galb," quoted in the chapter preceding this one, hints at nomads' traditional use of poetry as cartography. *Ḥassānīya*-language poets, like their compatriot Boisha and his fellow Friendship Generation poets, also make use of aeolian aesthetics. By doing so, they make visible Saharawi nomadic epistemologies. The latter are particular to nomadic lives in the Sahara, of course, but nevertheless share parallels with other Indigenous epistemologies. Bird Rose has described Indigenous knowledge as an "interactive knowledge in which people learn from the world around them, taking lessons in mutuality with the aim of keeping things flourishing in their country."[53] Saharawi *Ḥassānīya* poets' use of aeolian aesthetics, as with their compatriots who write in Spanish, serves to highlight Saharawi knowledge of the desert, to reject human exceptionalism, and to claim Western Sahara not because Saharawis own it and wish to control all its more-than-human components, but because they are *of* it. *Saharawi* in Arabic literally means "of the desert." Saharawi nomadic philosophies see nomads as one part of the desert ecology that interacts with, rather than attempts to control, other species living there, as well as with the weather, the water flows, the geological formations, and the aerosols. These latter coparts of the desert ecosystem impact upon and shape the lives of the nomads—sometimes helping, sometimes hindering. Either way, they are agents. Saharawi poets emphasize the agency of more-than-human parts of the windblown desert and its interaction with Saharawi nomads in a way that does not emphasize human control and domination. The inherent relationality between Saharawis—"of the desert"—and the more-than-human elements of the ecosystem complicates the human/nature binary sustained by the occupation's energoregime. The renewable energy battle therefore points to an ontological question as well as a political and ethical one. Insofar as nomadism is central to Saharawi national identity, are Saharawis still Saharawi if their entanglements with the rest of the desert ecosystem's entities are taken away? The focus on the intermingling between human and more-than-human in the poetry suggests that Saharawis can claim the Western Sahara because they *are* the Western Sahara, not just because they are Indigenous to it. Such existential questions are raised in the following poem "Tishuash," by the late El Badi:

Tishuash

(roughly "nostalgia," but perhaps more correctly, "the pleasure of remembering things that are passed")

All that has been has gone,
(how great the living and everlasting God!)
but how beautiful this scene is!

I see it sometimes—
no particular place—
just there with the goats,
like those nights I spent

at the mouth of a well,
making the wet sand my bed.
Enchanted by night's music:
the howl of wild dogs
and insects' whine.

Or in the watering season,
when the wheat is still to produce its seed,
I am there in the midst of the life of the camp,
doing some little thing about which you do not need to ask.

Or there we are
travelling in the dark before dawn,
from one stopping place to the next,
the only sound the swishing of camels' tails
before the sun has risen to our eye line;
walking on to those first lucid hours of the day
when the desert's features are clearest,
knowing both rocky valley
and the smooth.

And there again is the taste of tea,
flavoured with *da'i*,
in water sourced from the valley floor after the rains
or scooped from pools on concave rocks
where a river had run before;
when we were moving our camp from a dried-up well,
where the only firewood left was no better than kindling;
and I can smell that animal hide next to the spit
and see the clean bones beside that hide.

How come, my brother, you do not remember this;
the sweet life full of living?

It is no longer with us,
and if tishuash could bring it back
it would add tishuash
to the tishuash
of my tishuash.[54]

تشواش

مات الماضي سبحان الحي ◆◆ الدايم ماغلى منو زي

اعييت انشوفو ماهو في ◆◆ حاضر ما حاضر فيه أواش

كيف امبات اعل فم احسي ◆◆ بل اثراه اماسيه افراش

فالليل ينهول عندك عي ◆◆ لبهانيس وحس الخشاش

وللا يا للالك دهر الري ◆◆ زرعو ما فات اطلس لعراش

الفرݣان الحية واشوي ◆◆ ثاني ما عينك في التنباش

واسرية من حي الى حي ◆◆ فاتفاك امراكيبو تتناش

ݣبل اظهور الشمس اعل ري ◆◆ العين المعالم ف راش

تنظرها ما خافيك اودي ◆◆ منها فملاس ولا ف حراش

واتاي اعل مَ فيه الطي ◆◆ من تورطة وللا من مَ آش

ݣيݣ ابعيد اهلو من لحفي ◆◆ ماه اݣليل وحطبو ݣشݣاش

وافراش احذا بل الشي ◆◆ واعظام اعل ذاك افراش

عنك ما تتفݣد يا خي ◆◆ هذا من شي لذيذ ينعاش

ماهو ف الدهر اتل موجود ◆◆ لو كان التشواش اسوا لاش

نافعني فيه امال انعود ◆◆ زدت اعل تشواشي تشواش

Wind makes an appearance as an invisible agent here, in the wind-smoothed surfaces, the concave pools, the watering season, and the barley about to seed. It reminds us not only of nomads' deep knowledge of, and intricate relationship with, more-than-human desert existents but also of the *beauty* of this way of life. The translation by (Saharawi) Mohamed Sulaiman and (British) Sam Berkson retains words such as *da'i* (*ṭaī*) with the effect of emphasizing for the foreign reader the intricate relationship between the act of, and knowledge developed through, nomadism on one hand and the nomad's language on the other. *Ṭaī* is a *Ḥassānīya* word for the method by which nomads create a recipient for water using plants to line a hand-dug well.[55] Nomads dig such wells when they expect rain. But a key political moment of the poem (which occurs in the penultimate line in the *Ḥassānīya* original) is when the perspective slips to the second person. There is a question for you, "my brother." Up to this point, the pathos of the poem pushes the reader to share El Badi's sweet remembering (*tīšwāš*) of nomadic life. Enticing the reader into this poignant state of mind makes the turn to the second person all the more powerful. The question emphasizes the waning knowledge of nomadism and encourages Saharawi youth not to let this slip away. Nomadism is integral to Saharawi existence. Randi Irwin asserts that Saharawi identity, which is interlinked with a history of nomadism across Western Sahara, as well as the knowledge of the territory necessitated by nomadism, is prized as a form of evidence backing Saharawis' claims to Western Sahara beyond the current reliance on tribal membership.[56] For Irwin, such knowledge binds together Saharawi identity, territory (which, Irwin rightly argues, extends beyond earthly matter to the stars, clouds, and subterranean "territories" of Western Sahara), and history.[57]

Work by modern poet Malainin Lakhal likewise claims Western Sahara for Saharawis on the basis of love, care, and knowledge. Lakhal, who writes in Arabic, uses aeolian aesthetics to transmit his political messages. We know the wind through touch. We cannot see the wind, but we can feel its strength and temperature, and the sting, softness, wetness, and other sensations of the particles suspended in it. As explored in the previous chapter, Saharawi Friendship Generation poets use aeolian aesthetics to evoke the feel of the wind and windblown. As I discussed, Mohammed Salem conjures the desert touchscape, the feel of the windblown. She "walks barefoot on a rug of sand, soft as silk." Like Mohammed Salem, Lakhal evokes the feel of sand in his poem "Bedouin Obsession," an extract of which is below.

> She came to me with a memory of yearning
> With a bullet that her love placed
> In my heart
> And a tear
> Shed for both of us.
>
> She came to me with memories of the feel of sand
> Of her beauty when we made the desert our bed.
> Do you remember the colour of my eyes when the moon is full?
> Do you remember the poems
> You sang for our parting on those sleepless nights?
>
> ....
>
> The desert in my heart is a sea,
> And the sand is devotion.
> The clouds have names that flatter me,
> Night is a chain.
> The morning breeze that comes from the west
> Brings the memory of my love,
> And the longing for her joy
> Is an eastern blaze.
> The night is pain.[58]

جاءت تُذكِّرني ذكرى حنين
بنهدها رصاصة أودعها هواها
في القلب.
وبكت لعينيَ دمعتينْ
جاءت تُذكِّرني..
بالملمس الرملي..
بفتنتها ونحن نفترش الصحراء:
"هل تذكر لون عينيَّ،
حين ينتشي القمرْ؟
"هل تذكر الأشعارَ؟
غنيتُها لبعادنا، كي يحلو السهرْ.

. . . .

صحراء قلبي بحرٌ،
والرَّملُ عشقٌ.
والغيمُ أسماءٌ تغازلني،
والليلُ طوقٌ.
ونسيمُ صبحيَ في ذكرى الحبيبةِ
غربيٌّ،
والشوقُ لتفاح بهجتها
لهبٌ وشرقٌ، لهبٌ وشرقٌ،
والليلُ عشقٌ.

For Lakhal's poetic voice, the touch of sand is linked to memories of a beautiful night with a beloved woman. As Ellison remarks more generally, by personifying desert elements as lovers, Saharawi poets "affirm . . . affective commitment to the land."[59] As for Boisha in the previous chapter, gendered depictions of a desired woman allow Lakhal to express Saharawi love and care for the desert. Nods to the properties of sand, once again, underline Saharawi knowledge of territory. If one takes a handful of sand from the *bādīa* to a nomad in the refugee camps, she will be able to tell you the precise provenance of that sand from its look and feel.[60] Furthermore, the luxurious and sweet desert touchscape created with sand by Lakhal and Mohammed Salem is a contrast to Siemens's colonial construction of the Saharan winds as useless until "tamed" through their technical "innovations." Likewise, the idea of using sand as a luxurious bed (Lakhal) or a golden rug (Mohammed Salem) speaks of valuable and useful sands. The poets echo, here, nomadic forms of living with, and uses for, the medium of sand.

Anthropologist Sébastien Boulay highlights the several different types of sand known by nomads in the Sahara, as well as the various properties of sand that nomads value. Some sands are cherished above others when camping because, for example, they are more comfortable, supple, dry, and clean; they do not mark skin or clothes; or they are light enough to reveal any impurity on top of them, which can then be cleaned up or removed from the camp.[61] Sand might be used to perform a blessing before setting up camp, it might be the canvas on which to carve the name of the prophet in order to protect the camp's inhabitants, and it can be used to clean utensils or for performing ablutions.[62] Sand helps nomads to fix their tent in place and to cook bread and meat. It is part of nomads' cultural identity, argues Boulay, and affects their material cultures in the sense that the objects and utensils that nomads' fashion are adapted to the sand.[63] Beyond the camp, sand—material that is particularly receptive to external imprints—allows for tracking animals, other tribes, or—for purposes of navigation—the winds themselves.[64] Sand can entertain; it is used for playing (by carving grids into fine sand, for example, to create "board games").[65] Whether deposited on the ground or suspended in the winds as aeolian particles, sand is the "medium" in which desert nomads live.[66] By highlighting sand's useful and beautiful qualities, the poets analyzed here disrupt Siemens's colonial discourses on the medium and highlight Saharawi nomads' knowledge of, dependency on, and care for desert sands.

As in the patriotic traditions of many communities and literatures, Lakhal uses human romance as a metaphor for love of homeland. For Lakhal, talking of Western Sahara as a desired woman is a strategy to mask a nationalist message while under threat.[67] He penned the poem during a period of several months of hiding after having served two prison spells and receiving death threats for his pro-independence activism. As Lakhal told me, disguising his patriotic love of homeland with the memory of an encounter with a lover was a self-preservation strategy. However, Lakhal also genders the windblown desert to claim it not based on ownership but on love and knowledge. Like Boisha's poetic voice with his blemish or mole on a lover's belly, Lakhal's points to Saharawis' intimate knowledge of the desert by asking the imagined lover to recall the exact shade of the poetic voice's eyes when moonlight hits it. Winds are used to communicate the different emotions romantic love brings. Linking the "longing for her joy" with an "eastern blaze" (a reference to a the *īrīfī*) evokes passion along with empathy, while the morning westerly suggests gentleness. In its original Arabic, Lakhal makes liberal use of the breathy *ḥ* and *h*: this onomatopoeia is an oral and aural aeolian aesthetic expressing love. In the first stanza cited above, the original word used for *love* is *hawā*, which is both one of several Arabic words for (different stages of) love and a word for wind, or wind's blowing. The Saharan winds are tangled up with desire, for Lakhal. Western Sahara should be in the caring hands of those who love it, not those who would exploit it for wind energy.

Conclusion

The wind carries armfuls of sand, which it piles up against the walls of the adobe house. The air outside is pale and warm. In the living room, Gajmola, a primary school teacher, is reading a book to her five-year-old niece. We bought two copies of this book at the cultural festival. It is a bilingual book, Arabic/Spanish, published by Bubisher, a not-for-profit organization run by a team of Saharawis and Spaniards. I will read my copy to my sons. One day, Gajmola hopes to run a library.

Beside Gajmola, I am reading too. I find myself with Badi's "Tishuash" over and over and over. I cannot ever know what it is like to have been a nomad. But with the tangibility of the aeolian geomorphology, the specificity of nomadic vocabulary, and the attention he gives to each sense, Badi makes me feel something of what it must be like to miss such a life.

Traditional Saharawi nomadic poetry and other cultural forms that refer to the Saharawi nomadic way of life remind us that wind is a more-than-human agent.[68] Saharawi wind imaginaries are strikingly different from the colonial wind imaginaries of mariners, Spanish colonists, and modern-day energy multinationals and, I argue in the following chapter, can inspire an alternative, fairer energy future. Like the Saharawi Friendship Generation poets, Saharawi poets composing or writing in Arabic and *Ḥassānīya* use aeolian aesthetics to claim the Sahara for Saharawis on the basis of knowledge, love, and care. But wind is also used as a messenger and, linked to this, as an agent of solidarity—solidarity between Saharawis and Moroccans, solidarity between Saharawis and other desert existents, solidarity between Saharawis and humans elsewhere.

Part Four
Exile

CHAPTER 7

The *Gblīya*

Electricity in the Camps and a Nomadic Energy Future

I am in the camps in October 2019, at the end of the Algerian Hammada's summer. Moulud, the former nomad introduced in chapter 6, tells me it is the time of "the best wind," the *Gblīya*.[1] The *Gblīya* blows from the south. "It's gentle, good [and] lets people breathe."[2] When it comes, rain follows. It sets the feel of the season, directs peoples' emotions, and, in Moulud's words, "raises morale."

In the UK, two months earlier, the press reported a spectacular millions-strong influx of painted lady butterflies. Each year, the creatures stop in Europe to breed on their long journey from the desert fringes of North Africa. My son and I enjoyed counting the orange and black wings fluttering around the Buddleia in front of our house. It is a similar joy to see them back at their home in the camps, balanced on tree blossoms in the inner courtyard of a government building. I have just stepped out of a commemoration of the Saharawi Day of National Unity, held indoors.[3] In the center of the courtyard, there is a raised platform with a roof sitting on columns, like a rectangular bandstand. It has soft seating around its edges. I join a delegation from Andalusia that is resting there. It has been so hot these last few days. We quietly take pleasure in the breeze on our skin, the shade, and the papery shiver of leaves. A Saharawi guide jokes for the Spaniards, "There's some nice shade, but no cool beers."

Later, I tell Moulud about my day. I mention the pleasant air. "That was the *Gblīya*," he tells me. "They'll be some light rain soon." Sure enough, over the next few days, clouds form and raindrops fall—not enough to damage the adobe constructions, but sufficient to quench the thirst of the crops, and to refresh the mood. Yes, the *Gblīya* is a good wind.

Moulud is a father of four and grandfather of two. Before the exodus, he worked in the Nomadic Troops, the Indigenous unit of the Spanish-era desert police. He wore a uniform of billowing white robes and a blue turban and tracked "wanted persons" through the desert on camelback, often by night. "Wanted persons," he tells me, were principally Moroccan nationals perceived as threats by his Spanish superiors. The Spanish became suspicious of Moroccan immigrants following the Ifni War, explains Moulud. How did he navigate in darkness? The moon and stars of course, but also the plants ("a key way to know the wind is through its effect on vegetation") and the wind. For example, the *Sāḥlīya*, a coastal westerly, blows by night. If you feel the pressure of air after darkness, you know it is a *Sāḥlīya*. "You know which way is west."

The aim of this 2019 trip to the camps is threefold. I wish to learn how the energy system here developed, why it grew in the way that it has, and how people use it. I wish to learn enough about nomadic uses of the wind to be able to understand Saharawi wind imaginaries, especially as expressed in poetry and other forms of cultural production. Obviously what constitutes "enough" or "understand" is debatable. I will never know *enough* about Saharawi nomadism, or Saharawi culture more generally, to fully *understand* Saharawi poetry. To rephrase, then, the second ambition of my fieldwork is to learn all that is possible to learn during the limited time period of my fieldwork, and within the limits of my position as a non-Saharawi, non-refugee, white, sedentary, privileged, British academic.

At the time of this fieldwork trip, which is just over a year ahead of the November 2020 return to armed warfare, Polisario has been making exploratory plans for a return of the refugee population to the liberated territories. An important part of such scoping concerns how to meet the energy needs of a returned population. A third aim of my fieldwork trip is therefore to explore the planned future energy system of a free Western Sahara in the liberated zone.

In this chapter, I present the findings of this fieldwork trip in conversation with the findings of my earlier poetry analysis. In other words, I aim to establish to what extent the existing and planned energy systems of the SADR-in-exile and a liberated SADR in the free zone, respectively, are built on and reproduce Saharawi wind imaginaries. These imaginaries, as shown in part 3, are in turn rooted in wider Indigenous Saharawi, culture, thought, and knowledge. This chapter therefore furthers my argument that different wind imaginaries inform energy systems in specific ways, that wind imaginaries are a central component of (renewable) energy systems, and that Indigenous wind imaginaries can potentially shape a fairer energy future.

There are links between Saharawi culture-informed infrastructure and the infrastructures of Indigenous communities elsewhere. As Jennifer Wenzel argues, through what she calls "resource logics," humans understand nature as something other to themselves, existing for the use of humans, and subject to humans' control.[4] Ann Spice calls the infrastructures "developed" according to such resource logics "invasive infrastructures."[5] These infrastructures cement and further capitalist exchange, settler land ownership, and colonial dispossession. Spice contrasts "invasive infrastructures" with the "Indigenous critical infrastructures" of Indigenous peoples of Canada. The latter are the interconnected networks and cycles that link humans and more-than-humans, and which make Indigenous life and survival possible. For Spice and her Indigenous research partners, networks consist of living organisms and inorganic substances—the salmon and berries that humans eat, the bears that pass the berry seeds, the berry-bearing plants, the water that feeds them—and the relations between them.[6] These are relations that need care from all members of the infrastructure.

If, to use Spice's terminology, the renewable developments in occupied Western Sahara are "invasive infrastructures," what would an "Indigenous critical infrastructure" of energy look like? Jordan B. Kinder, using the case of the Lubicon Cree Nation (an Indigenous community in what is now commonly called Canada), argues that Indigenous models of solar energy infrastructure are premised on maintaining good relations between humans and more-than-humans.[7] "Indigenous solarities" are, says Kinder, solar energy systems built on Indigenous epistemologies and ontologies and are a "form of solidarity between the human and nonhuman world."[8] Kinder's vision relies on Spice's expanded sense of infrastructure and on "alternative imaginaries," which focus on relationships between organisms and matter.[9] While settler colonialism and its energy systems undermine Indigenous communities' land-connected practices and epistemologies, Indigenous solarity, on the other hand, facilitates such practices.[10] I argue in this chapter that Kinder's emphasis on the need for a renewable energy system to facilitate land-connected practices, epistemologies, and solidarities between the human and nonhuman worlds resonates for Saharawi nomads. Below, I explore the extent to which energy developments (projected and actual) in the Saharawi state-in-exile constitute an Indigenous critical infrastructure that relies on Saharawi nomadic (wind) imaginaries and epistemologies and facilitates land- and wind-connected practices.

In the *Ḫaīma*

On the first morning of my 2019 trip, Embarka[11] brings juice and dates from the kitchen. Her mother tells her to go back for the other dates, "the better ones." The latter, once fetched, join boiled eggs, cheese triangles, baguettes, honey, olive oil, and fruit on the table. Abundance is hospitality. Maimina[12] pours the coffee. It tastes of berries. "It's from Kenya," says Maimina. She worked there, in an SADR government role, before she was married. Life is quite different now. Her husband works abroad and sends money back. The daily cooking, cleaning, tea brewing, shopping, and washing, as well as caring for her small children and elderly relatives, are tasks that start before dawn and end well after nightfall. Such tasks need energy—all sorts of energy. Hosting guests must be an additional burden, although she never lets me feel that I am such. Maimina heads one of the more well-off families, relative to other refugee-citizens. Electrical appliances make the household work easier and quicker, and she even has enough money spare to rent an air-conditioned apartment in Tindouf during the scorching summer months.

Temperatures in the Algerian hamada reach far higher in summer, and dip lower in winter, than in milder Western Sahara. In the heat, sandstorms hurtle towards the camps and engulf them, like yellow tsunamis or orange Niagara Falls. The suspended sand participles filter the sunlight until the world glows crimson. "It's like an apocalypse," says Maimina's brother Limam.[13] He shows me a photograph—"no filter"—of a flame-red sky on his mobile phone.

Flash floods are also a problem. The climate crisis has brought long droughts interspersed with short but intense periods of rainfall to the Algerian Hamada.[14] Each time such floods drench the camps, dozens of thousands of refugee-citizens are left homeless. Sometimes there are cholera outbreaks.[15] While nomads had ways to predict and deal with the scorching easterly winds and storms, the sedentary refugees—far from their homeland and the place-based solutions and mitigations it offers—are vulnerable. Houses in the camps are made with mud bricks for several reasons: using only local materials means self-sufficiency, only human energy is needed for building, refugee-citizens can earn a living making the bricks, and such materials are temporary, thus maintaining the hope of a return to the homeland.[16] But mud bricks crumble to nothing in floods. Moulud recalls for me the last big flood, when his house collapsed: "We all had to live and sleep on a hill until it was dry enough to come back and repair the damage."

The Saharawi community is a textbook case in climate injustice. The Saharawi contribution to climate crisis is negligible and yet the Saharawi refugee-citizens face the harshest symptoms of the crisis. The destructive capitalist Global Northern lifestyle has provoked, and continues to worsen, climate change.[17] Meanwhile, top-down climate politics undermine ecologically sustainable ways of life such as nomadism, and corporations take over official international processes such as the United Nations Framework Convention on Climate Change (UNFCCC).[18] This is starkly illustrated in occupied Western Sahara, where climate finance furthers climate colonialism. The refugee camps are situated on the boundary of zones of the "high" and "extreme" risk categories for heat exposure.[19] Nevertheless, the SADR is denied a voice at global climate change talks because it is not a UN member state. This means it is unable to apply for the technical and financial support mechanisms for climate change mitigation efforts that are available to other states.[20] Meanwhile, as Nick Brooks shows, Morocco uses these same mechanisms (in 2019 alone, it received $293.8 million in formal climate finance) to position itself as a global leader in the struggle against the climate crisis. Brooks eloquently explains the implications of this asymmetry in support mechanisms:

> Morocco has increased the exposure and vulnerability of the Sahrawis by occupying their homeland, displacing most of the population, denying them access to their own natural and economic resources, forcing them to live in areas that have the fewest resources and that are most exposed to the impacts of climate change, preventing them from accessing external technical and financial assistance as a result of the unresolved conflict, and marginalising those Sahrawi living under occupation. Yet Morocco receives large amounts of climate finance, while the SADR receives nothing. In this way, the system of global climate governance and finance favours one actor in a conflict while disadvantaging the other. . . . This is the very definition of climate injustice—systems, actions and finance intended to address climate change that instead systematically reinforce the structures and power relations that drive vulnerability.[21]

After breakfast, Limam and Gajmola show me their solar panel and how it connects to their battery. "El Aaiún camp isn't connected to the grid yet." At the time of this visit in autumn 2019 (and indeed the visit before that in winter 2015), the pro-green-energy government is in the midst of connecting the camps to the fossil-fueled Algerian grid.

At the Ministry

The SADR's energy department, part of the Ministry of Transport and Energy, opened a few years ago, and energy infrastructure is still not fully under its mandate. Like most of the republic's government institutions and NGOs, it is based in Rabouni camp. The Saharawi state-in-exile comprises six camps. Five are residential, and four of these (El Aaiún, Smara, Auserd, and Boujdour) are within a one-hour drive of the Algerian town and military base of Tindouf. The fifth, Dakhla, is some eight hours away by Land Rover. The sixth camp, Rabouni, holds ministries, other government bodies, media organizations, and NGOs. Several Saharawi workers also temporarily reside here, as do some foreign visitors. To serve the transient workforce are restaurants selling pizza, goat kebabs, and rice dishes. But Rabouni, like the other camps, also produces foodstuffs. Not far from the ministry, Saharawis have grown a date palm oasis from seed. Its name, Nkheila, is from the Arabic word for *palm tree*. It is one of several small farms that refugee-citizens developed in the 1970s and 80s and which is watered using drip irrigation networks.[22] There are also small greenhouses: onion, tomatoes, carrots, and leafy vegetables grow in rows under white domes supported by bamboo structures and insulated with palm fronds. The produce is destined for the state-in-exile's hospitals and most vulnerable families. Camels, kept for their milk, are penned within a circle of old car skeletons.

I walk over the sand—crispy after the rain—to the ministry. Step, crunch. Step, crunch. As always, I take my shoes off at the door. Like almost all buildings in the camps, the ministry is a single-story adobe construction built around an inner courtyard. Mohammed Saleh, one of the managers of the energy department, ushers me in and gives me some water. We sit on armchairs that are still covered in transparent plastic. Soon, Mohammed Saleh is joined by the department's director as well as the secretary-general of the Ministry of Transport and Energy, also Mohammed. The latter sits behind a desk with a small SADR flag. I introduce myself, my research project, and my volunteering with Resource Watch and other groups. Mohammed jokes, "If your organization is not successful in stopping Morocco stealing resources, we'll have no choice but to keep applying for funding for our energy projects here in the camps."[23]

After the 1975 exodus, Polisario ensured each family had a gas cylinder for their energy needs.[24] A Swiss humanitarian organization brought the first solar installation to Smara camp in 1987. "The idea was initially to see if solar energy would be viable in the camps," explains the secretary-general Mohammed. Further collaborations with NGOs brought more solar energy resources in the coming years. "We don't like asking for help, but we are obliged to." Mohammed elaborates, "We rely on a policy which is developing human resources. At the moment we are focusing on educating and training the population, to generate specialists in every kind of field. Because this is the only thing we can work on and improve without our own finances. Then when we get our independence, we will have people that can develop things. . . . Now we have people who can manage all aspects of renewable energy, but we need further finance to realize our scoped-out projects in the free zone and in the camps."[25]

Given the SADR's limited financial resources, when foreign support for solar energy systems began to arrive in the 1980s, it had to prioritize. Pharmacies, schools, and women's training centers were top of the list to receive panels.[26] The priorities for solar power connection were (and are) identical to the wider national priorities of the SADR government, which proudly publicizes its achievements in three key areas internationally: the sophisticated health system that it has developed in the camps despite scarce resources, its attainment of the highest literacy rates in Africa, and its alleged achievement of gender equality.[27] This will not be a surprise to scholars of energy politics. Energy infrastructures have played a significant role in processes of nation building internationally.[28] Such infrastructures convey imaginaries of the state and are commonly associated with state and nation building.[29] From France to Russia, energy systems are tied up with notions of national identity.[30] Likewise, Marcus Power and Joshua Kirshner argue that the study of electrical infrastructure in Africa can reveal how states seek to "read" their territories or citizens.[31] SADR nationalist discourses, I have argued elsewhere, see gender equality as an intrinsic component of Saharawi society.[32] Women's prominent position in Saharawi society, according to official SADR discourse, was reversed by Spanish colonialism, which left Saharawi women without educational and economic opportunities or political or social standing. A key goal of the Saharawi nationalist revolution, then, has been to return women to their "traditional" and "rightful" position in society, including by offering them education and employment opportunities. Most women's training centers were originally located at a dedicated camp (27th February camp, now called *Boujdour*), where women could live with their children, with free daycare while they studied. The fact that the women's training centers were among the first public institutions to be electrified illustrates how energy policy follows nationalist discourse on the need to promote gender equality.

At first then, the SADR was a solar-powered nation. From the late 1980s until 2015, decentralized solar power dominated the camps. Public services (hospitals, schools, media centers, and so on) were powered by small solar photovoltaic (PV) systems. Some cultural centers had (and have) a mixture of solar PV and wind turbine power. Most families had a 150-watt solar panel or two to meet their energy needs at home, sent by relatives or friends abroad or, latterly, bought from the market.

In 2015, connections to the Algerian grid began. Why? The time of neither peace nor war (1991–2020) saw mass migration out of the camps. Refugee-citizens sought sources of income in Mauritania and Spain. In response, Polisario tried to improve the material well-being of refugee-citizens that stayed in the camps. They did so via infrastructural developments, including roads and improved health, education, and social facilities, as well as symbolic salaries for its civil servants.[33] Grid connection is an extension of this government departure from their imagined, "ideal" nation in favor of responding to the demands of (some of) its citizens. Mohammed Saleh further explains, "Foreign NGOs helped initially by donating solar panel systems, but when these break beyond repair, long after the NGOs have left, what are we to do?"[34]

Refugee-citizen opinions on the grid are split. There are some refugee-citizens who oppose further grid connections on the same grounds for which they prefer using mud bricks for buildings rather than more permanent materials: the camps are temporary, and thus, all structures should likewise remain as portable as possible. Then Saharawis can pack up and go home to Western Sahara "once" they achieve independence. However, some refugee-citizens that support grid connections do so on the grounds of comfort. As one interlocutor put it, "We have been here over forty-five years. I hope we'll return to Western Sahara eventually, but in the meantime, we might as well be comfortable." Relatedly, Dunlap points out that the reasons why people desire ecologically destructive development are complex.[35] Who could possibly deny the right of a refugee-citizen—who might have only an old unrepairable solar panel donated by the charity of the former colonizers—to be connected to an electricity grid that they hope will be more reliable? During a field visit to Auserd camp in December 2015, the popularity of grid connections was noticeable. My host family lived on the outskirts of the camp, and almost all their neighbors' homes were empty and abandoned. They had moved their lives to another camp where there was already a grid connection. During my 2019 trip, Limam tells me, "When the grid technicians come to the *ḥaī*, women go out on the streets and cheer."

Several refugee-citizens argue that solar panels could be just as reliable as a grid connection if they are well looked-after. They also argue that a small energy system with a few panels could provide more than enough energy for a family. However, several families contend that their panels do not provide sufficient energy or that they do not have funds for repairing or replacing old panels (the ministry civil servants tell me that only refugee-citizens with the means to afford it pay for grid energy, thus subsiding the poorer members of the population, although the ministry admits that this causes complications). The SADR's decision to connect to the Algerian grid in order to provide a more reliable connection for refugee-citizens can be considered as a gendered one since domestic and caring duties fall disproportionately on women. Having electric light at home gives the ability to turn on a television to entertain the children, power a fan to cool the vulnerable family members in the excruciatingly hot summers, recharge a mobile phone to call a husband working abroad, play the radio to hear news from the frontline. The SADR's decision to begin connections to the Algerian grid is one that arguably benefits women to a greater degree, at least in the short term (the long-term effects of continued reliance on fossil fuels is, of course, gendered and racialized).

Mohammed, Mohammed Saleh, and their other colleagues at the ministry would all personally prefer to rely exclusively on renewable energy. However, they do not have the funding that they need to provide a solar system for each family or the funding to maintain such systems. All ministry officials and workers that I speak to see a future independent Western Sahara as running on renewable energy. Mohammed explains, "Renewable energy is both cheap and valuable at the same time. For solar energy, we have the sun, and we also have wind, because we are in the desert. So this is a good opportunity to invest in renewable power. One has the satisfaction of relying on oneself, and not depending on other things. . . . In any case, the desert belongs only to God, and at the same time belongs to everyone who loves it."[36]

The last line of this quotation expresses a sentiment and wording that I frequently hear from refugee-citizens old enough to have known nomadic life. Curiously, it is likewise a perspective heard by Howe among some of her *istmeño* research participants: "It belongs to God or no one," they told her of the wind.[37] Such perspectives are also reflected in Saharawi poetry. As I argued in part 3, Saharawis claim Western Sahara not on the capitalist basis of ownership but on the basis of their love and care for all the desert elements—human and otherwise.

The SADR government's official position on an energy future for a free Western Sahara is similar to that of Ministry of Transport and Energy staff. Connecting to the Tindouf grid has enabled "greater energy access" for the refugee citizens, yet the SADR maintains its aspiration to return exclusively to renewable resources if it were to receive the necessary financial support. To quote the SADR government's first indicative Nationally Determined Contribution (iNDC) to the Paris Climate talks, "Expansion of renewables to adequately meet the energy needs of the Sahrawi refugee population, with external support via climate finance, would support both the energy independence of the refugees and help to reduce emissions in the near-term. The expansion of small-scale renewables would also provide an energy source that is readily transferable to and scalable in the territory of the SADR once the decolonisation process has been completed."[38]

We should be wary of tendencies to conflate Indigenous cultures. However, aspects that several cultures identifying as Indigenous have in common include, firstly, place-based ways of life, which are necessarily acutely environmentally aware and, secondly, ongoing, lived experience of colonial and/or settler colonial oppression. Both aspects shape communities' attitudes to energy infrastructure. Writing of Indigenous communities in the territory today known as Canada, Leela Viswanathan argues that a key motivator for Indigenous-owned-and-operated renewable projects is energy autonomy. She conceptualizes energy autonomy as a form of self-determination. In her words, energy autonomy "reflects a community's ability to generate, store, distribute, and sustain an energy system locally, without the need of external intervention."[39] Of course, energy infrastructure in the liberated zone is planned to enable and incentivize a return of the refugees to Polisario-controlled "free" Western Sahara. In this way, small-scale renewable energy developments are the building blocks of a physical Saharawi staked claim to Western Sahara. Again, such politicized understandings of self-sufficiency are common to other Indigenous populations. As Viswanathan points out, some Indigenous communities in North America participate in renewable energy developments with the purpose of asserting "their collective rights to land and self-determination."[40] For the SADR, renewable energy is the path to energy autonomy, which in turn means independence (in the political, national sense) and self-sufficiency in the sense that it would prevent reliance on foreign players.

What happens to the old solar panels from those camps that are now connected to the Tindouf grid? Mohammed Saleh explains, "In the *wilāyāt* that aren't electrified, like El Aaiún, they replace broken panels by recycling panels from other *wilāyāt*."[41] This policy of reuse is replicated at the family level and in the market economy of the camps, which began to emerge in the 1990s during the aforementioned period of "no peace, no war."[42] During the first war, the socialist and nomadic principles of the Polisario were lived in practice. All were equal, there was no money. Everyone applied their skills to the national effort as best they could: men on the front line and women building hospitals and schools, nursing the ill, and teaching children. When the era of "no peace, no war" began with the 1991 cease-fire, people migrated and sent back remittances, and some started businesses. Poorer and richer families emerged.

Many Saharawis working in Spain saved up to send a solar panel home. Others imported panels to sell.

In the *Marsha*

Laayoune market is the biggest in all the camps. Turbaned men wearing flowing garments sit outside shop facades. Small groups of women wander slowly and upright along the sandy thoroughfares, their colourful *āmlāḥaf* fluttering in the wind. Every inch of their skin is covered from the sun and blowing sand. Gloves, scarves tied into turbans, thick socks, and sunglasses block out what the *melḥfa* does not. One should be careful which shoes one wears in the warmer months—the heat of the sand can melt soles.

Gajmola, Limam, and I go into Hartan's store. The floor is mosaiced and spotless. It smells of soap. Hartan sells suitcases, *āmlāḥaf, darrāʿa*, bras, sets of matching bags and shoes, long-sleeved dresses, rugs of weaved plastic, shower gel, children's clothes, a brand of face-whitening creams called *Moon Face*, shampoos called *Funny* and *Fancy*, and solar panels. He buys the panels secondhand from Mauritania. The secondhand ones were manufactured in Europe. He also sells new ones, sourced from Mauritania but manufactured in China. He tells me that they are cheaper in both senses of the word.

Europe, Mauritania, China. The decentralized renewable energy system in the camps reaches out across borders and even across continents. This has important implications for energy, environmental justice, and for a so-called just transition. For Saker El Nour, cofounder of the North Africa and Middle East Network for a Just Transition (RÉSEAU TANMO), a just transition should entail a local participatory approach.[43] It should also focus on the achievement of autonomy from the former colonizers, reducing poverty, and mitigating the effects of climate change and environmental degradation. According to Jawal Moustakbal, who works on the Moroccan case, a just transition must focus on local sovereignty and solidarity at every stage of the energy system.[44] Likewise, for Myles Lennon, who works on the case of black communities in the Global North, a just transition should be community-led, accountable, and publicly controlled.[45] However, Lennon does point our attention to the politics of supply webs. While the solar PV used in the camps do not rely on fossil fuels to the extent that concentrated solar power (CSP) plants do, it is still worth reflecting on the global environmental justice implications of solar PV, which relies on extraction for some of its materials.[46] Some Chinese solar companies reportedly use forced labor of ethnic minorities.[47] As Lennon highlights, it is currently impossible to completely extricate renewable energy from systems of racialized and classed inequality.[48] These are systems that privilege predominantly white and rich investors and consumers over communities where resources are extracted and whose labor is exploited.[49]

Of course, corporate, colonial, large-scale developments fueling the oppressive energy system in occupied Western Sahara are forms of racialized environmental injustices associated with so-called renewable energy. Nevertheless, even single solar PV panels used to power a lightbulb or telephone recharger in a refugee's tent cannot escape this racialized, capitalist system. Decentralized, refugee-led renewable developments in the camps are at the opposite end of the scale and, for this reason, I have not preceded the word *renewable* with *so-called* while discussing the Indigenous small-scale infrastructure described in this chapter. Nevertheless, the supply web behind solar panels illustrates that a fully "just" transition is impossible while the capitalist system still exists. For Lennon, the solution is "supply chain solidarity," which involves advocates for decentralized, community-led solar projects "forming alliances with activists and workers in the various industries that manufacture solar infrastructure to better identify their exposure to co-pollutants and other exploitative labor practices, and their demands."[50] For Dunlap, the root to "real" renewable energy (he conceptualizes the corporate-led so-called transition to be nothing more than fossil fuel+, a continuity of fossil fuel regimes) is through degrowth measures as well as ensuring the proximity of energy production and consumption, energy self-sufficiency, and the communing of energy infrastructure.[51] Degrowth aims at the abolition of economic growth as an objective of society and the reduction in human consumption of natural resources. It has been defined as the "voluntary transition towards a just, participatory, and ecologically sustainable society."[52] The need to degrow is relative, of course, and is much more urgent in the high-consuming societies of the Global North. The traditional Saharawi nomadic economy was never one aimed at growth. Saharawi wind imaginaries, as explored in part 3, emphasize the interdependence and solidarity that exists between the human and more-than-human existents of the desert. Furthermore, and as discussed above, the role of Saharawi refugee-citizens in provoking the climate crisis is negligible. Small-scale domestic and community solar and wind systems to provide energy for refugees match Dunlap's requirement of the communing of energy infrastructure and are explicitly aimed at self-sufficiency even if necessary materials are not (yet) produced locally.[53]

Abdelhay's shop is three doors down from Hartan's. It smells of smoking coal and green tea and incense, a haze of which lingers in the air. Abdelhay only sells the cheaper, new solar panels from China. He makes the journey to Mauritania twice a year—once more than Hartan. It is a hard trip. It is mostly off-road, and it takes two weeks there and back. But it is all done "above board"—the merchants get "papers" from the Polisario, and they all pay a small customs duty to the Mauritanian authorities. Another shopkeeper sells only the better-quality, more expensive, second-hand panels. The next one I speak to has stopped selling altogether; since some camps have been connected to the Algerian grid, demand for panels is falling.

Of the families connected to the grid, some keep the panels for use in case of grid power outages. Others sell them to families living in the camps that are yet to be connected to the grid—Smara, most of El Aaiún—bypassing the market altogether. Reuse, repair, and refurbish reigns.

Shop *Light*

El Wali's electrical repairs shop is on the outskirts of the *wilāya*, by a main road. It is a small pristine-white building with solar panels leaning neatly against its outside walls. It contrasts aesthetically with its next-door neighbor: a much larger car mechanic's-come-welder's yard and workshop, displaying a Mad Max-esque sandy showroom of repurposed rusty cars. I recall the back catalogue of poet El Khadra, her odes to the *Boyevaya Mashina Pekhoty* (BMP, the USSR's first mass-produced infantry fighting vehicle), and the late poet Beibuh's "The Land Rover," or *draīmīza* (hornless goat), as it is nicknamed by Saharawis.[54] Polisario refurbished and armed both for battle.

In front of the two businesses is a large rectangle of sand reserved as a car park, which is marked out by evenly spaced half-moons of rubber tires. The yard's entrance showcases the resident tradesman's skill by way of a huge arch, like a blue whale's jawbone, of welded scraps of metal. Al Wali's shop is announced by a more modest, but just as beautiful, painting of the sun rising behind three desert *gallāba*. Above the painting is the shop's name in Arabic calligraphy: "Shop 'Light:' the Repair of Electronic Devices."

El Wali looks up and smiles, but the pliers in his hands do not stop working as he greets us. Two women wait while he repairs their power inverter. He will see us when he is finished their job—it won't be long. We three go outside where I take advantage of the time to elaborate my field diary. Gajmola, a talented photographer, takes shots of me in the same frame as the solar panels—Gajmola rightly points out that the researcher should be documented just as she documents her research participants. We look at the photos together. The women leave, and El Wali calls us in.

Work is just starting to quiet down after a busy summer. The hot months mark the time of broken solar panels, thermostats, power inverters, and cables. "That's all anyone brings," says El Wali.[55] As well as the shop, he has been working in the public energy sector for ten years. Like hundreds of Saharawis, he trained in Cuba, spending his entire adolescence, age twelve up until his early twenties, in the Caribbean. Now he has the shop to make a living and simultaneously volunteers his time on behalf of the nation at the Ministry of Transport and Energy's energy department.

El Wali has two fields of responsibility in the energy department. On the one hand, he works as a member of a team of technicians to ensure the smooth rollout of the networked electrical infrastructure in the camps. His particular role involves liaising with his Algerian counterparts to find solutions to difficulties, iron out any problems, and resolve any miscommunications—an energy infrastructure diplomat, in a way. His other field of work looks to the future. He and three other Saharawi electrical engineers are preparing a government-requested scoping study for a renewable energy future in the liberated zone. As Porges has highlighted, the SADR government has led and realized a land mine eradication program over several years with the precise objective of facilitating nomadic pastoralism in the liberated territory.[56] El Wali and his team are tasked with calculating how to meet the energy needs of communal institutions and nomad families there. They take stock of existing infrastructure—there are already wind-powered communal wells for the nomadic families that currently roam in free Western Sahara—while planning what additional wind and solar infrastructure will be necessary to sustain the welfare of a much larger population. The SADR government's energy strategy for the liberated territory will rely completely on renewable energy. It would extend this 100-percent-renewable-energy strategy to the territory currently occupied by Morocco upon decolonization. The government envisages decentralized, small-scale renewable energy, agricultural, and water infrastructure. Such a model is seen as best suited to supporting nomadic pastoralism, especially as heat extremes grow with climate change. As the SADR government explains in its first indicative NDC, "Small-scale agriculture based on decentralised renewable energy, as is being piloted in the Liberated Territory, may be more sustainable than the industrial scale export-driven agriculture promoted by Morocco in the Occupied Territory. . . . An end to the conflict, the dismantling of the Berm, the clearing of munitions, and the development of small scale (e.g. water) infrastructure in the Liberated Territory and the areas currently occupied by Morocco, would enable the rehabilitation of resilient, traditional pastoral livelihoods."[57]

Usually, the liberated territory supports a small population of nomadic families, reportedly up to sixty thousand persons before the November 2020 war.[58] There are some public institutions in the liberated territory, such as

Tifariti hospital and several administrative offices and pharmacies, all of which currently depend on solar energy.[59] A Swiss NGO, the Association pour le Développement des Energies Renouvelables Sahara-Solaire-Solidaire (founded in the 1980s), is currently working in partnership with Saharawi engineers to develop an autonomous photovoltaic system for use in remote locations. This would deliver partial air-conditioning in health facilities and electricity for lighting, refrigeration, and medical equipment. The Swiss/Saharawi team is also examining the potential for solar thermal sterilization of medical instruments.[60]

In terms of water sources, there are today no surface water reserves, and, as discussed in chapter 3, the military occupation combined with climate change are increasing Western Sahara's thirst. Following rainfall, which Saharawis can use the winds to forecast, nomads traditionally hand dig wells in locations known for groundwater. More recently, and in response to the drying up of the land, with financing from another Swiss NGO (Association Secours Sahraoui, established in 2011), Saharawi engineers have built solar PV-powered wells in the liberated territory. Two are in the region of Bir Lehlu—completed in 2018 and 2019—and produce forty cubic meters of water each per day, and one is in Tifariti, completed in 2020, and produces seventy-nine cubic meters a day.[61] This may not sound like much for all readers of this book—in a country like Spain, adults use on average two hundred liters of water a day—but Saharawis might have a different perspective. Refugee-citizens in the camps must make do with, on average, ten liters—equivalent to 0.01 cubic meters—per person per day.[62]

When it comes to residential needs, El Wali and his team (working on behalf of the SADR government) quickly discounted the option of large-scale solar farms. They questioned whether such a model would be "adaptable enough"—adaptable enough to last a long time in the local weather, and adaptable enough for nomadic lifestyles. In the end, the team decided to scope the possibility of providing every Saharawi family (that is, every tent, each of which is normally headed by a woman, who might have a husband, children, parents, and siblings living with her) with their own portable, independent solar energy technologies: a nomadic residential energy system. El Wali explains, "We find that families in the liberated zone often travel, so it's good if they have their independent panel, that they can transport, have their own independent network if you like."[63]

The Saharawi wind imaginaries explored in part 3 of this book emerge from nomads' various interactions with, knowledge of, and reliance on the winds and the windblown. Saharan winds have shaped and enabled nomadic life.

They facilitate navigation, forewarn nomads of dangers, bring food and forage through seed dispersion and cloud-suspended rain, and—depending on the specific wind or windblown phenomena—lift the spirits and inspire poetry and song. Winds are a basis for nomadic existence. Traditionally, Saharawis used heat energy in the form of campfires (always made from deadwood) and human energy to dig wells, tend to plants, and care for camels, goats, and other humans. Their practices and philosophies have significant parallels with energy uses among other communities. Diana-Abasi Ibanga analyzes and compares the corpus of African Indigenous philosophies to arrive at an Indigenous, African energy ethics. His conceptualization incorporates ethical worldviews that are common to all the philosophies he studies. Ibanga highlights the Annang maxim of caring for the land on which one depends; the Ubuntu ontologies of the interdependence of humans, other animals, plants, and inanimate beings; and Ibuanyidanda philosophical inspirations on the rightness of sharing one's living space with other existents.[64] Ibanga also highlights the Annang belief that no existent, human or otherwise, can claim sole ownership of land, which he compares to the Braai ethic of sharing space and resources not only with more-than-humans but also with humans and other existents of the future.[65] Ibanga points to Braai, the practice of Indigenous South Africans that involves bringing people together around a fire (i.e., around energy, as Ibanga points out) to share space and resources.[66] Without conflating these diverse Indigenous philosophies, he pulls out their parallels and commonalities in order to formulate an African conceptualization of energy, which views energy as a force in which all existents participate and as a force that holds the universe in balance.[67] Retaining this balance and sharing this force therefore requires complementarity and relating to other natural forms rather than exploiting them.[68] In practical terms, to apply African energy ethics to renewable energy developments means accommodating, and showing gratitude to, nonhumans and future (non)humans in all decisions, restoring any losses caused by the developments, and avoiding any negative implications for (future) existents.[69]

The small-scale renewable energy systems proposed for the liberated territory through El Wali's team depart from Saharawi traditional energy sources of human and (cherished) animal labor, and the heat of a fire. Yet they also emerge from, and are vital for maintaining, the windblown, nomadic way of life in the face of growing climate crisis challenges. For example, hand-dug wells are becoming a decreasingly reliable form of accessing water due to increasing droughts. Wind-powered wells offer a solution for nomads without undermining nomads' traditional care for their ecosystems or damaging their relationships of reciprocity with desert other-than-humans. To use the vocabulary of Spice and Kinder, this is an "Indigenous critical infrastructure" of energy. El Wali's team's plans, and existing renewable energy infrastructure in the liberated territory, are also arguably in line with Ibanga's conceptualization of an "African energy ethics." Renewable energy developments are planned in solidarity with other existents and with a future nomadic population in the liberated territory.

The occupation and new war threaten this nomadic energy future in obvious and direct ways. The destruction of infrastructure. The displacement of nomads from the liberated territory. The slaughter of people, plants, and other animals. Ecocide. Slow violence against nomadism. But the longevity of the occupation and war also make indirect threats. My aforementioned friend Hamza, an English literature graduate, often jokes that the Saharawis are "waiting for Godot." Their three-decades wait for the international community to make good on their 1991 promise for self-determination broke Saharawis' patience, resulting in the 2020 war. It also put pressure on the SADR government to connect the camps to the Algerian grid, ending the former's reliance on small-scale solar and wind developments. The longed-for independence has also led to policies incongruous with the Ministry of Transport and Energy's envisaged nomadic energy future.

The Petroleum and Mines Authority (PMA)

While the SADR government commits to a renewable energy future *for Saharawis* in a free Western Sahara, its PMA has also committed to selling Western Sahara's potential oil resources to foreign corporations in the case of independence. This would undermine several of the nomadic principles of SADR's own ministry's plans for the liberated territory. The sell-off of oil blocks to foreign companies, after all, contradicts the inter-existent solidarity at the heart of the envisaged nomadic energy future and constitutes instead what Lohmann has, as outlined in the introduction to this book, described as "white energy"—a conceptualization of energy rooted in imperialism and capitalism—or Ibanga's conceptualization of "Western energy": human exploitation of other natural forms for energy without due care to the well-being of those energy sources.[70] How can we comprehend such a radically incongruous pair of policies as those of the SADR Ministry of Transport and Energy and those of the SADR PMA?

The Saharawi Republic's PMA has responsibility for granting petroleum exploration permits. At the time of writing, it has agreements with four international companies for eight permit areas. The agreements, *assurance agreements*, should convert to production-sharing contracts following any UN Resolution that assures the immediate realization of Saharawi self-determination. The Saharawi Constitution allocates authority to government for the administration and stewardship of resources in the national interest and asserts that natural resources (including mineral riches, energy resources, and resources of the seabed and territorial waters) are "the property of the people."[71] The SADR claims to have consulted extensively with civil society ahead of launching its licensing round. However, the agreements are contested by some Saharawis. Some of my (younger) interlocutors oppose extraction of petroleum and minerals on environmental grounds.[72] On the other hand, Randi Irwin's research among Saharawi young people in the camps found groups that supported the PMA's assurance agreements because they aligned with youth activism against oil companies entering into agreements with Morocco without the consent of Saharawis. But Irwin also interviewed youths that were critical of the SADR and other Arab countries for relying on oil when solar energy was in such abundance.[73]

We should take into account how, and why, the SADR presents the PMA's activities externally in order to understand the drivers of policies such as petroleum assurance agreements. Kamal Fadel, Saharawi diplomat and official of the PMA, consistently refers to the concerns of the international community about the stability of a free Western Sahara and simultaneously presents the assurance agreements as part of a wider SADR natural resources policy to ensure "the development of a viable, self-reliant and democratic nation which will contribute to peace, stability and progress of the Maghreb region."[74] A key argument of Morocco and its allies in favor of Moroccan sovereignty over Western Sahara is that the "International Community" cannot risk a failed SADR-led state in Western Sahara. Several scholars have pointed out that the SADR is under constant pressure to conform to the Global North's expectations—or, one could say, blackmail—in return for aid and the hope of independence.[75] This includes proving its democratic credentials, its "moderate" interpretation of Islam, the liberated position of Saharawi women, its dedication to ensuring economic and political "stability" in the Maghreb post-independence, and its intention to provide opportunities for trade and "development." As Mark Drury argues, the SADR has had to engage in a "a decades-long struggle to make [its] camps legible to the international community (and, largely, the West) as not only camps, but as components of a viable nation-state."[76] Drury understands this enforced show as a component of the "international politics of recognition."[77] In other words, the SADR does not have the luxury of ignoring the Global North's perceptions of its future economic "viability." Such pressure should be considered among the drivers of the Saharawi Republic's wider energy policy.

As other drivers, the SADR explicitly states that the strategy behind the petroleum licensing is twofold: "to deter Morocco's efforts to exploit the country's natural resources and to prepare for the recovery of full sovereignty."[78] Irwin's cautious conclusion is that such agreements can indeed serve decolonization efforts. To quote Irwin, "In the case of the Western Sahara, the implementation of the deferred oil contracts has generated new social, political, and economic forms that create a political terrain, transforming otherwise inaccessible, seemingly off-limits, resources into productive spaces of opportunity in the present through their very projection of a possible future. Here, Saharawis have found a way to create new forms of territorial control essential for their political future."[79]

In the pursuit of self-determination and decolonization, the PMA's petroleum agreements are in line with the aims of the SADR's energy department. However, they also mark a radical departure from the nomadic and wider African energy ethics that I outline above, sacrificing solidarity with more-than-humans for the price of nominal independence. It would be easy for someone like me, with no experience of forced exile, war, or brutal oppression to criticize such a stance. But we cannot forget the "International Community's" actions, which constantly force the SADR to jump through hoop after hoop in pursuit of Saharawis' right to independence. Policymakers, politicians, and academics, myself included, talk of the integral importance—in terms of the planet's future—of learning from Indigenous knowledge. The quickest route to losing that knowledge is by preventing Indigenous peoples themselves from practicing it. This is the case for Saharawis, for most of whom—despite SADR programs such as summer schools in the liberated territory to initiate young people in the basic skills of nomadic pastoralism—nomadism is now a memory.[80] The tactics of the PMA should be read in this context of urgency. Indeed, Maria Stephan and Jacob Mundy have argued that these assurance agreements were initially an SADR strategy to provoke a legal battle between competing oil companies (those with agreements with Morocco versus those that hold agreements signed by the SADR), which would help SADR achieve political independence for Western Sahara.[81] Of course, it is entirely possible that the PMA's assurance agreements are *nothing but* a tactic or strategy, that there is no intention behind them to ever sell off oil blocks upon independence, and that they are *exclusively* an attempt to—as Irwin puts it—project a possible independent future for Western Sahara. If this is the case, the incongruity between the Ministry of Transport and Energy and the PMA would not be so incongruous at all.

Passive Energy and Human Energy

When Taleb Brahim was a young boy, he fled to the refugee camps in the Algerian hamada, where, as his five-year-old self soon found, "no grass grew."[82] He is now an agricultural engineer, trained through SADR's solidarity higher education programs with Libyan and Syrian governments. Taleb lives with his family on the outskirts of Smara camp, still far away from the electrical grid. He greets me in a billowing *darrāʿa*. We walk together, talking about Taleb's job, to the small farm behind his house.

Taleb's work focuses on developing small-scale agricultural innovations such as low-tech hydroponics units. Hydroponics is a type of horticulture that involves growing plants without soil. By *low-tech*, Taleb refers to technologies that make use of local materials readily available to, and affordable for, refugee-citizens. The units recycle water and use naturally produced fertilizers. Taleb was driven by "sustainability, self-sufficiency and independence for Saharawis."[83]

On the farm, I can hear water trickling somewhere. It is a modest, underground reservoir, which irrigates the whole farm, and through which water is recycled. A single 150-watt solar panel powers the pump. The farm is divided into patches. In one patch, there is small raised pool, tiled blue. Trays packed with pepper, tomato, and aubergine plants float on the surface of the pool water. Elsewhere, fish swim and feed on algae. Their dung acts as fertilizer for the plants. Another patch is covered with a fabric dome. Inside there is shelf upon shelf of abundant green barley, like spiky green carpets. The barley receives nutrients through solutions rather than soil, thereby using far less water and growing twice as fast as it would if using traditional forms of agriculture. "Also," explains Taleb, "the grains of barley liberate some nutrients in the water which are good for vegetables." I can smell mint: "Mint absorbs fish urine, so it acts like a biofilter. Fish need their water to be clean." Butterflies and dragonflies flutter and zoom around the farm. "Everyone helps everyone. The system meets its own needs."[84]

This "sustainable self-sufficiency," to use policy-speak, emerges from nomadic traditions. "Saharawi nomads used to solve any kind of difficulty that we faced, but we needed to move to do so. When there was a lack of grass or any sort of food for the animals, we followed the clouds." In a situation of forced sedentarization, Taleb Brahim and others draw on the traditional adaptability of nomadism but find new solutions rooted in self-sufficiency, independence, and a mindfulness of other members of the desert ecosystem. At the edge of the farm, there is a row of potted aloe vera. "They're for medicinal purposes," comments Taleb. "I hope one day every Saharawi will have one."[85] At the time of writing, around 1,200 Saharawis in the camps have replicated Taleb's low-tech hydroponics systems to create their own home gardens.[86] As Porges puts it, describing the sedentarized refugees, "[Their] nomadic mentality persists in the form of opportunism, flexibility, social solidarity and resisting single points of economic failure."[87] As far as Taleb knows, he is the first person globally to develop low-tech hydroponics in conditions that are widely considered to be extreme in terms of climate and availability of resources. With Taleb's help, the World Food Programme (WFP) is now trialing his model in seven other countries with large refugee populations and extreme conditions.

Taleb's current work began with wishing that his fellow refugees could enjoy their traditional kitchen. The Saharawi refugee's basket of rations from the international community dwindles annually and is nutritionally insufficient, lacking in fruit, vegetables, and Saharawis' traditional staples of meat and milk. "If we talk about food, it's about pleasure. To enjoy what you are eating—that has its culture." How to grow fruit and vegetables in the rocky Algerian Hamada? And how to grow fodder to sustain camels and their precious and nutritional milk? The usual agricultural solutions rely on infrastructure and resources that are not available to the refugee-citizens: large quantities of water, chemical fertilizers, pesticides, and stores for hay. In any case, as Taleb says, "If you insist that pesticides and fertilizers are necessary for agriculture, then you will rely on multinationals."[88]

Taleb dreams of moving to Mhiriz, in the liberated territory of Western Sahara. As Joanne Clark and Nick Brooks point out, the inland regions of Western Sahara are more humid than the rest of the Saharan territory and Atlantic zone.[89] Taleb of Mhiriz says, "There are many resources there. There's dew every morning, and there are low-tech ways of harvesting that water. When it rains, it rains a lot in a short amount of time. How to store that? We could look to other Indigenous peoples' traditions."[90] He would like to develop small-scale self-sufficient means for Saharawis to grow plentiful fodder for their camels in the liberated territory, just as he has done in the camps, but using only human energy and simple, low-tech, passive energy. Taleb explains, "Saharawis in the liberated territory get no food aid, they rely completely on their animals. And they can't get to our traditional grazing lands because of the berm." For humans, there are already small farms across the liberated territory. Food production there is community-led, small-scale, and geared toward self-sufficiency and independence. Community gardens are clustered around

Tifariti, Aguenit, Mhiriz, Zug, and Bir-Lehlu and provide for the hospital at Tifariti as well as local schools. Some families also grow home gardens. They eat well: aubergine, garlic, onion, courgette, carrots, turnips, beetroot, melon, lettuce, spinach, Swiss chard, parsley, mint, and coriander. Gardens with greenhouses add pepper and tomato to the dinner menu. Pomegranate, fig, and date palm trees provide dessert. Labor is manual. Saharawis in the liberated territory use no pesticides or chemical fertilizers. Flood irrigation or small-drip irrigation are used.[91] The SADR government would like to expand such home gardens and small farms with a view to encouraging oasis-like microclimates, as they have managed in the camps (through the gradual creation of the aforementioned Nkhaila oasis in Rabouni camp, for example).[92]

The development of small-scale farms or community gardens in the liberated territory is entirely within Saharawi tradition. However, conditions are far more difficult than in the past due to climate change and the negative ecological impact of the occupation—hence Taleb's desire to develop low-tech innovations. Historically, several Saharawi *qabā'el* practiced seasonal or itinerant agriculture, growing, for example, barley in *grāra* (*agrāīr* in plural). *Grāra* is a *Ḥassānīya* word for an oval-shaped topographical depression whose main axis aligns with the direction of the dominant wind.[93] These depressed areas had the advantage of relatively humid clay or mud substrata, which held nutrients. The deepest parts of *agrāīr* would also become watering holes for nomads' *ẓāīya* (a word to denote pooled rainwater) animals after rains.[94] Traditionally, nomads' sedentary agricultural practices were limited to oasis areas where they planted date palms, which, through care, they restored and grew.[95] Taleb's innovations allow Saharawis to continue the traditional nomadic preoccupation with ecological restoration in a time of climate crisis in line with the above-discussed conceptions of "Indigenous critical infrastructure" and "African energy ethics."

Conclusion

Researchers have questioned whether the Western imaginary of centralized management of energy systems is achievable in the Global South.[96] The Saharawi nomadic energy imaginary questions whether it is desirable. Small-scale renewable and agricultural developments, both those that exist and those that are planned by El Wali's team, Taleb Brahim, and others, emerge from, and maintain, a Saharawi nomadic ontology. These nomadic ways of being and knowing share parallels with several other Indigenous communities in their emphasis on the solidarity between the various existents of the desert ecosystem and the rejection of human ownership of the land. Saharawi small-scale, potentially portable energy and food systems fit into Spice's conceptualization of an "Indigenous critical infrastructure" and Ibanga's "African energy ethics." They also aim for Saharawi independence and self-sufficiency with a focus on the use of affordable local materials. This local focus shows solidarity with the current victims of the racist, capitalist supply webs upon which "renewable" energy too often relies. It also shows the nature of an energy system that Indigenous wind imaginaries inspire.

But this nomadic energy future is under threat. The international community's conflict irresolution and other policies are pushing the SADR to leave these energy systems behind. Rob Nixon, drawing on the works and life of Nigerian anti-oil extraction and Ogoni rights activist Ken Saro-Wiwa, argues that environmental self-determination is indispensable to the cultural survival of Indigenous peoples.[97] This is true for Saharawis. But so too is the converse: cultural survival depends on self-determination. Forty-five years of occupation have made nomadism impossible for the vast majority of Saharawis. Most of the SADR-in-exile's population was born in the refugee camps and do not know how to live nomadically. Estranged from their nomadic culture, Saharawis' demands and desires for a decentralized renewable energy system are displaced in favor of increased comfort in very harsh material, economic, political, and spiritual conditions. In part due to the coloniality of climate finance schemes, SADR cannot afford to ensure that each refugee family has a functioning domestic solar energy system in the camps. But it can ensure domestic grid connections in which better-off refugee-citizens subsidize the cost of energy for poorer families. Meanwhile, the SADR government, forced by a hypocritical international community intent on appeasing Morocco, has entered into the resources game, allegedly (one cannot be sure of the SADR's intentions behind the scenes) promising oil contracts upon the achievement of Western Sahara's independence.

Indigenous knowledge is not something that can be extracted from the Sahara, learned from, and applied in the Global North. Saharawi Indigenous knowledge is a practice tied to nomadic culture, windblown desert ecosystems, and specific wind imaginaries. With decades-long military occupation, exile, and injustice, Indigenous knowledge fades away, and the nomad-poets die—and with them, potentially, the nomadic energy future.

CHAPTER 8

The Sirocco

Saharan Winds and Solidarity in Spanish Writing and Art

No plant can grow without phosphorus. Every year, the dense green jungles of the Amazon lose around twenty-two thousand tons of phosphorus to rain and flooding. Obligingly, every year, the Saharan winds, specifically the *īrīfi*, take some 27.2 million tons of dust and sand, containing an estimated twenty-two thousand tons of phosphorus, on a transatlantic journey to the world's largest tropical rainforest.[1] Twenty-two thousand tons. Just the right amount to replace what was lost. An organic solidarity.[2]

Wind moves in a borderless world. It traverses continents, oceans, and cultures. Yet, by relegating wind to the background, we ignore its multiple social and political possibilities. In this chapter, I build on chapter 6 to foreground the potential of wind imaginaries in reflecting and engendering solidarity, but I move from Saharawi poetry to modern Spanish art and writing. I am interested, particularly, in the nature of modern Spanish imaginaries of Saharan winds and how these link to the solidarity necessary for a fairer energy future. Researchers have devoted entire monographs to defining the concept of solidarity. For Sally J. Scholz, political solidarity is "a moral relation that marks a social movement wherein individuals have committed to positive duties in response to a perceived injustice."[3] Solidarity has its own history in different parts of the world, including in Western Sahara. Polisario leaders and founders of the Saharawi Arab Democratic Republic (SADR) were heavily influenced, ideologically, by Fanonian Pan-Africanist currents as well as left-wing, anti-colonialist movements of the sixties.[4] Franz Fanon's articulation of Pan-Africanism was a call for unity between North Africans and Sub-Saharan Africans in the context of Western hegemony and the pursuit of freedom from colonialism.[5] It was a solidarity forged around Africans' pursuit of liberation from colonial powers—a solidarity free of national borders. For Fanon, national liberation would be meaningless if a neighboring country was still under colonial rule.[6] SADR was a founding member of the organization that is today called the *African Union*, and, for that very reason, disgruntled Morocco remained the only African country not to be a member until 2017. The SADR continues to pursue political solidarity with fellow African nations. Most of SADR's embassies are in Africa and Latin America.

The SADR, and Saharawis more generally, also pursue ties of solidarity with Spain, albeit not necessarily ties inspired by the Pan-Africanist tradition. Scholz's description of popular understandings of solidarity is useful for conceptualizing Spanish-Saharawi solidarity. She notes that "solidarity" is often used to signal friendship, to unify, to stand in for sympathy, and to rally social protest.[7] In a poem cited in my introduction to this book, Hamza Lakhal asks the wind why it has not carried the Saharawis' message abroad. He implies that Saharawis expected the wind to disseminate calls for solidarity. This chapter looks to Spain for an answer, focusing on the entwinement of Saharan wind imaginaries and Spanish-Saharawi solidarity in Spanish creative output.

I selected material for this chapter in a semi-serendipitous way. Originally, I envisaged that the chapter would focus on modern Spanish novels set in Spanish Sahara. I therefore selected works by Felix Romeo, Javier Reverte, and Jesús Torbado based on their provocative engagements with Saharan winds. Nevertheless, I stumbled upon Francisco Bens's memoir and Ramón Mayrata's autobiographic prologue to his edited collection of nonfiction writings on Spanish Sahara upon researching chapter 2. I found Alberto Vázquez-Figueroa's and Ramón Mayrata's novels when reading Mahan Ellison's book *Africa in the Contemporary Spanish Novel, 1990–2010*. I came across Federico Guzman's audiovisual installation *The City of Wind* when he shared a video of the animation on a social media channel. It was through the installation that I first encountered Fatma Galia Mohammed Salem's poem of the same name, which I analyzed in chapter 5. During a research trip to the Saharawi camps, I happened upon the children's book *Sand/Water* at a book exhibition by the Saharawi/Spanish cultural project Bubisher. I use the installation and children's story in this chapter due to their interesting and deep engagements with Saharan wind.

Windblown Nostalgia

A handful of ex-residents of Spanish Sahara have written memoirs, or at least short life-writing texts, to share their memories and reflections of their time there. Several of such texts are by retired military personnel—unsurprising since most of the peninsula-born population of the Sahara comprised soldiers and their families. In his memoir written well into his retirement, Francisco Bens, the once living legend of the desert, aches for Spanish Sahara. Wind is his vehicle for communicating feelings of nostalgia. Although the memoir was published in 1947, over twenty years after he left his post in Spanish Sahara, Bens recalls his return to Spain with a deep and raw longing for the desert. As if culture-shocked, he feels as if he can no longer "fit in" in Madrid. The *īrīfi* is the symbol he chooses to represent how his move from Spanish Sahara to the metropolis burdens him: "I had changed the *īrīfi*, the incredibly powerful wind of the Sahara that makes mountains of sand change place, for the air of a ventilator! . . . I was no longer meant to be a city man. Madrid was still and stale and, in the Sahara, even the sand is a traveller."[8]

Desert wind is also the motif of choice for Ramón Mayrata when expressing his own feelings of difference in a new place and culture. Mayrata first traveled to Spanish Sahara in 1973, tasked with assisting Saharawis in writing a history of the country for use in the schools of a future independent Western Sahara. That book was never finished—those schools are not yet opened. Mayrata did, however, write a novel based on these experiences (see below) and edited a collection of Spanish colonial writings on Spanish Sahara, first published in 2005. His prologue to this collection starts and ends with pleasantly disorientating winds. The opening paragraph reads, "At first the mind feels disconcerted. The wind transforms everything that is not itself. The desert flows and changes shape, disguised in an eternal monotony of sand."[9] The prologue concerns Mayrata's first visit to the territory, how mysterious the desert seemed to him and other visitors, and how foreign he felt. Significantly, he affectionately gives the wind a "self," a soul. By doing so, like Bens, Mayrata conveys a sense of nostalgic longing to the reader. He closes his prologue with Spanish Sahara rendered exotic, affectionately other, and again uses the wind symbolically to do so: "In spite of the difficulties of this people, each dawn is still a complicit gesture from the horizon, to stretch and leave in pursuit of a landscape that dissolves in the wind."[10]

Spanish author Alberto Vázquez-Figueroa's memoir *Sand and Wind* is likewise marked with nostalgia and loss.[11] It focuses on Vázquez-Figueroa's adolescence spent in Spanish Sahara, which, he tells the reader, was the "the most wonderful period of [his] existence."[12] For Vázquez-Figueroa, the desert "will always be the sign of happiness . . . , that place where, for others, there is only heat and thirst, fatigue and sun, sand and wind."[13] The memoir is titled *Sand and Wind* because "there is something that humans can't fight against: nature.

And [in Spanish Sahara] nature has two names: sand and wind."[14] Spanish Sahara's sand and wind are both objects of, and vehicles to express, his nostalgia for the desert and his happy times there. In his prologue, Vázquez-Figueroa writes of his "nostalgia for the great plains, the wind that cries on the dead shrubs, and the legends told around the love of a campfire."[15] He reminds the reader of the West's typical terra nullius-informed imaginary of the Sahara Desert—"uninhabited, without lifeforms"—and spends the book undoing such a limited understanding of desert. He fills the Western-constructed "emptiness" with Spanish Sahara's "true" fullness and beauty.[16]

Deserts have long held a cherished place in Orientalist hearts. From European and North American twentieth-century literary subgenres of desert romance with rapacious sheikhs and enticing harems, to the fantasies of colonial heroism of modern-day "desert tourists," to ideas of romantic wastelands, constructions of the exotic and mysterious desert have been central to Orientalism.[17] Evocations of local winds, although not well researched, contribute to such constructions of Eastern deserts. Spanish writers certainly evoke the Saharan winds with romantic feelings. However, Mahan Ellison has convincingly argued that, despite the tendency to exoticize the Saharan territory, several modern (post-1975) Spanish authors generally write in ways that undermine the historical power of Orientalism.[18] Likewise, Nizar Tajditi says, "The dream of the desert still conserves its charm and attractiveness for new Mediterranean writers. However, it has turned into an open dream, that is, a dream that seeks to settle into the dreams of the Sahara's inhabitants and coexist with them in a wide cultural and human framework of imaginaries. Here, there is no discrimination between superior and inferior ethnicities."[19]

There is certainly affection for, but no intentional solidarity with, the Saharawis in Vázquez-Figueroa's and Bens's memoirs. First published in 1961 and 1947, respectively, the books are a product of the colonial period. Nevertheless, their use of Saharan winds to evoke deep feelings of nostalgia for their lost happiness in the desert is reminiscent of the techniques of Saharawi writers. There is an (unconscious) effect of producing empathy in readers, who might begin to reflect on what it feels like to miss the lost Saharawi homeland. Mayrata, however, has long been and continues to be an active member of the Spanish movement of solidarity with the Saharawis. His writings put him in the category of Spanish authors writing on the Sahara post-1975, which as Conx Moya points out, "try to assimilate the desires of the Saharawis themselves, and are written from a position of non-superiority. It is often a literature of strong political and emotional commitment with the Saharawi people, with a bad conscience."[20]

I would go further and argue that feelings of bad conscience in the modern Spanish novel on the Sahara evoke a deeper sense of betrayal, a betrayal that demands reparations in the form of committed political solidarity, which are explored respectively through metaphors of blood ties and wind.

Betrayal, Blood, and the Blowing Wind

Ellison describes Mayrata's novel *El imperio desierto* as in pursuit of "coexisting, equally valued differences" and as "a powerful counter-Orientalist work of fiction."[21] As Ellison points out, the title of the novel could be translated as *The Desert Empire* or *The Deserted Empire*.[22] The latter of the two fits better with the content of the novel, which tells of the final months of Spanish colonialism in the Sahara and, crucially, of Spain's betrayal of the Saharawis, selling them off to Morocco and Mauritania. First published in 1992, the novel follows the character of Ignacio Aguirre, a Spanish anthropologist who, just like Mayrata himself (and therefore suggesting that the novel is semi-autobiographical), is sent to the Spanish Sahara to write a history of the Saharawis, which is to be taught in schools post-independence. However, as the political situation develops, and Morocco and Mauritania stake a claim to Spanish Sahara, Ignacio's mission changes. The school textbook is forgotten. Instead, Ignacio works with Saharawis to collect evidence—oral and written documentation that shows their ethnic, historical, and cultural differences to Moroccans and Mauritanians—in favor of Saharawi independence for the International Court of Justice (ICJ) tribunal. Spanish authorities create barriers for Ignacio and his Saharawi colleagues throughout this endeavor, including kidnapping and incarcerating Ignacio, but he eventually compiles eight thousand pages of documentation and hands it to the government in Madrid for the consideration of the ICJ. The novel ends as Ignacio departs El Aaiún for the last time, as the Spanish population, who are unable to look the Saharawis in the eye, flee and as Morocco and Mauritania invade. The novel is one of betrayal but is written from a position of solidarity with the Saharawi cause, tracing, through the characters and the oral histories they collect, the various arguments in favor of independent statehood for the Saharawis.

Wind features prominently in the novel. It works as a plot engine by foreshadowing events to come. For example, wind acts as a harbinger as Ignacio first leaves for Spanish Sahara. Mayrata introduces the scene of Ignacio's airport departure from the Canary Islands to El Aaiún with three paragraphs that describe the effect of the *īrīfi* in Las Palmas. The heat is so intense that it seems like "the stains of oil and gasoline on the runways are about to explode into flames," it is "difficult to breathe" and "a cloud of very fine dust" floats in the air, causing a fellow passenger to cough. "Without doubt, something's going on in the Sahara," concludes Ignacio, the wind seemingly the root of his suspicions.[23] Of course, the very presence of the *īrīfī* in the Canaries is significant. Named moving masses of air cause a shared physical sensation (of heat, of the touch of sand on the skin, of being pushed by an invisible force) between Saharawis and Canarians, reflecting their entwined history and present.

Toward the end of the novel, Mayrata's use of the wind as a foreshadower comes full circle: the reader finds themself in the vicinity of an airport once again as Ignacio leaves the Sahara. Ignacio's last view of the land below is one of whirlwinds of sand "provoked by wind," a "wind that challenged the inscrutable figure of commander García Ramos."[24] As García Ramos is presented in the novel as the chief Spanish supporter of handing over the Sahara to Morocco, the whirlwinds daring him to a duel suggest future Saharawi resistance.

For Saharawi characters in the story, the wind is a messenger, or a metaphor for one. Persuading fellow Saharawis to share oral histories with Ignacio, Saharawi General Mohammed Fadel explains, "When he writes the truth, the whispers of our elders' voices will spread throughout the world like the wind and no one will doubt that we should be an independent people."[25] Similarly, when Saharawi Sidi Azman gives books and documents from an ancient Saharawi library at Smara to Ignacio, he explains, "I hope that they serve to proclaim to the four winds that we are an independent people."[26]

Mayrata also uses wind to set the scene in more general terms, giving the reader a sense of the Saharan land-, sound-, and touchscapes. These wind-centered 'scapes challenge terra nullius narratives of the desert. For example, at a meeting between Spanish officials and Saharawis in El Aaiún, "a smooth breeze ascend[s] from the dry bed of the Saguia el Hamra making the leaves of the trees in the garden jingle slightly."[27] Evoking a garden of rustling leaves immediately undoes the received idea of desert-as-wasteland in a Western reader's mind, and a "smooth breeze" seems familiar and tender, as opposed to the gothic, howling desert wind of Siemens's mind's ear.

Novelist and journalist Javier Reverte (1944–2020) also takes Spain's abandonment of the Sahara as the subject matter in his novel *El médico de Ifni*. The story follows Clara, who is on a quest to find out more about her late estranged father Gerardo, who (like many real-life peninsula- and Canaries-born residents of Spanish Sahara) joined Polisario fighters when, in his words, Spain "betrayed" the Saharawis.[28] What she finds, as Ellison notes, leads her to embrace the Saharawi cause and look upon her Spanish relatives with disgust.[29] In the course of the novel, she discovers that her father had an affair with Saharawi Fatma, by whom he fathered a child, Omar: Clara has a Saharawi half-brother. Clara travels to the Saharawi refugee camps in search of Omar, and, after learning that their father was betrayed, Clara and Omar together decide to avenge him.

Throughout the novel, wind is used to provoke empathy in the reader: it acts as a vehicle to tell the reader of the emotions of the characters. Alexandra Harris points out that, across cultures, we find external images to represent inner feelings or ideas: inner emotion becomes outer weather.[30] In *El médico de Ifni*, wind establishes the emotional economy of the novel by reflecting and thus communicating to the reader the otherwise unspoken feelings of the characters. When Clara almost has an erotic encounter with a young man she meets in the street, the hot, breathless emotional and embodied anticipation of the event is reflected in the air, which becomes "thick" and "lacks oxygen" before thinning and becoming cooler when the two would-be lovers part.[31] Similarly, when Clara and Omar learn of the role of the slippery spy Alberto Balaguer in the death of their father and decide they want him dead, a terrifying *Gatma*—a *Ḥassānīya* word for dark (black, red, or grey) and blinding sandstorms that appear incredibly rapidly—suddenly descends "as if springing from a giant invisible hand that blows a dirty rug with a broom, lifting unstoppable whirlwinds of dust from the ground."[32] A final example comes from Gerardo's diary, in which he indicates his indifference to death: "I write in my *jaima*, once again I feel alone while, outside, an earthy wind and blinding sandstorm whips, called the *liaayaa* here. It's an uncomfortable and unhealthy wind; but I take pleasure in it, because it numbs the senses, the soul and the ability to reason. If the *liaayaa* always blew, perhaps there wouldn't be wars. And maybe not love either, or loyalties or even guilty or innocent men."[33]

The wind directs an inward eye to Gerardo's condition of depression, evident in his desire to lose the ability to feel emotion. If solidarity is a form of empathy or sympathy, of being able to feel another's emotion, then, in *El médico de Ifni*, wind is its vehicle.

Wind plays other roles. It is the subject of several of Gerardo's poems, which pepper the novel and add pathos to the scenes of Clara reading the diary of her troubled late father. In one such poem, Gerardo's lover Fatma is described as "coming from the wind," reflecting either the common Spanish colonial conflation of Saharawis with the wild desert winds or, more likely, Saharawis' own poetic view of themselves as of the wind.[34] In the words of Hamza Lakhal (quoted in chapter 6), "From the wind [we] come / To the wind [we] walk." Wind is used to echo Saharawi sentiments in other ways. Just as Boisha does in his aforementioned poem "Say That You Haven't Told Me," the place of Saharawi exile is compared unfavorably to Western Sahara's lost heartlands. To create this contrast, the winds of the Algerian hamada are imagined differently from those of Western Sahara. The former winds are nasty, destructive, and violent. They hurl rubbish about the camps and carry an unpleasant smell. The winds in Spanish Sahara, however, often come in the form of pleasant breezes, or as strong, artful winds, sculpting dunes. Suelma, Clara's Saharawi lover, likewise echoes wider Saharawi voices as communicated in Saharawi poetry, claiming the desert on the grounds of love, respect, and knowledge. Suelma tells Clara, "No one will take the desert from us anymore because only we understand it. The desert itself will expel those that today occupy it: because they don't understand it and have not learnt to love it."[35] If solidarity is to raise the voices of those who have less platform, then, again, wind is the vehicle of Reverte's solidarity in *El médico de Ifni*.

There is less solidarity for Saharawis in Jesús Torbado's *El imperio de arena*, which offers sympathy instead to the Spanish citizens forced to leave their homes in Spanish Sahara and the former Spanish protectorate of Sidi Ifni. Through flashbacks and flashforwards, the reader relives the final forty years of the Spanish protectorate in Ifni and colony in Western Sahara. We follow the story of Elisa, who arrives as a youth during the Spanish Civil War to seek refuge with her aunt and uncle in Sidi Ifni. We also learn that Elisa's parents had a second motivation in sending their daughter to the Sahara: they believed (in line with comments in earlier chapters on colonialism and meteorological pathology) that a warm climate would cure her chest problems. In the present moment of the story, elderly Elisa, who stays behind after Spain's 1975 exit, is trying to recover some property, which, she says, some Moroccans have stolen from her. Following her pursuit for justice throughout the novel, we simultaneously learn of Elisa's tragic life story. Wind brings her downfall: her fiancé died in a storm days before their wedding, and she gives away their baby to avoid a scandal. At the end of the novel, we discover that the properties are not Elisa's after all. Unbeknownst to her, her uncle had gambled them away years before. Nevertheless, Elisa remains in Ifni at the close of the novel, hoping for the return of her long-lost son. Betrayal and disappointment are key themes in the novel, echoing the duplicity and betrayals of Franco and the Spanish government toward Spanish citizens of the Sahara, argues David Ross Gerling.[36] Both Elisa's son and her property, not stolen by Moroccans but given away by her uncle, can be read as metaphors for Spain's handing over of Ifni and the Sahara to Morocco. The construction of the novel, in which Elisa's story is told parallel to a history of the political intrigues that led to the Green March, add credence to such a reading.

As in Reverte's novel, wind builds the emotional architecture of the story, changing its mood to reflect events and its actions mirroring what characters would do to each other. When the youthful Elisa has a heated discussion with her aunt, "outside, there were frequent gusts of wind, and the window shutters shook."[37] We also come across a word common in vernacular Canarian: *Asirocado*. Elisa describes the Arab maids that mock her as "asirocadas," thereby using wind to convey the mental landscape (or windscape) of certain characters.[38] The Sirocco is a name used in the Canaries and elsewhere in the European Mediterranean for hot, humid, sand-bearing winds blowing in from the Sahara. Some Canarians and southern Andalusians accuse the Sirocco of causing headaches, insomnia, and depression. In the Canary Islands, "asirocado" also refers to the time of the Sirocco, the season of its blowing. We might imagine the dictionary entry: "Asirocado: 1) a person worn out or driven mad by the Sirocco wind; 2) the time period in which the Sirocco wind is usually prevalent."

That a named Saharan wind has inspired its own vocabulary reveals the extent to which it is tangled up with daily life and culture in the Canary Islands. Elsa López's ethnography of the Sirocco in the Canary Islands finds that, for her informants, the Sirocco is "the personification of the enemy; something real and tangible that forms part of their lives and that can be directly responsible for certain misfortunes that affect their lifecycles and work."[39] Resentment toward hot winds, blown at the Canaries from the Sahara's scorching lips, also appears in Mayrata's novel, as the words of a fellow passenger of Ignacio's reveal in a scene at Las Palmas airport: "The irifi, the cursed wind from the Sahara's interior," comments a passer-by as Ignacio struggles to breathe in the heat and dust.[40] To describe the abstract notions of anticolonial resistance, or the colonizer's fear of the Indigenous Other, wind, yet again proves the most evocative metaphor.

Just as, according to López, the Sirocco and *īrīfi* are enemies of the Canarians, the *īrīfi* is also the nemesis of a central character in Felix Romeo's *Discotheque* (2001). Romeo (1968–2011), a writer, translator, and literary critic, claimed that his novel was based on anecdotes that he heard while serving twelve months in prison for refusing to perform either military or social service (this was compulsory for all Spanish men until 2001). *Discotheque* is an aeolian tragedy in which wind brings the drama. It is written from several different perspectives. The voices that we hear most are those of Torosantos, an ex-boxer turned sex show actor, whose professional and romantic partner is another protagonist; Dalila Love, a black, transgender woman with a mysterious past; and Torosantos's unnamed father, an Ifni War veteran obsessed with recalling sinister memories of his time in the Sahara. Torosantos's father has gambled away his son's life. We follow his quest to find Torosantos in order to kill him. We also follow the lives of tragic heros Torosantos and Dalila, who garner the reader's sympathies with their kind hearts and multiple misfortunes. The novel shocks. It is speckled with bestiality, incest, child rape, sexual exploitation, torture, and murder, mostly committed at the hands of Spanish legionnaires all of whom are simultaneously struggling against, or recalling their struggles against, "moors."

In the novel, the *īrīfi*, a blast furnace of a wind, is simultaneously Romeo's muse, a metaphor for all the worst fears of the Spanish legionnaire and a tool to mock Spanish imperial patriotism—Romeo's satirical flourishes storm around in gusts. Other winds appear as supporting characters—the *aliseo* (trade winds), for example—but the *īrīfi* is part of the novel's scaffolding. It is an agent that moves characters to act in a way that propels the story. And its metaphorical role foregrounds the colonial fear of indigenous resistance. An early chapter, written in the voice of Torosantos's father and entitled "How the Thirteenth Battalion of the Legion Disappeared in the War of Ifni" features a hot wind "bringing the song of the moors, starting in Arabic but the hot wind translated it."[41] The song concerns a snake that will "give you death" and is followed by allusions to the Green March.[42] Throughout the novel, the "Moors" continue to "fuck [the Spanish army] over with the wind and dust."[43] For instance, the "Moors"' wind drives a soldier mad, prompting him to kill his compatriots.[44] Romeo builds, here, on a tangible fear well established in the colonial psyche and discussed in previous chapters, from the days of shipwrecked sailors through the Spanish colonial period in the Sahara, of the desert's uncontrollable, indomitable winds.

The *īrīfi* continues to be an agent throughout the novel, passing messages to the legionnaire characters and increasing their paranoia and fear, usually in association with their conflict with "moors." For example, the wind tells Torosantos's father that he will be killed.[45] Later, a chapter entitled "Something Else about the Story of Torosantos' Father about His War in Ifni, and How He Tells It to Mercedes Ibarra, the Owner of the Guesthouse in Las Vegas" is dedicated to describing the *īrīfi*, "the most terrible wind that you know, more terrible than a hurricane, much more terrible," and its effects.[46] Romeo exhausts the meteorological tongue in his chapter-long analysis of this desert wind. The description of the *īrīfi* includes its visual effects—how it transforms the landscape and impedes visibility, and its work as "the most marvelous of sculptors" forming filigree and hamadas of rock; its auditory effects: how it sings, how the sound of blowing sand is like a whistle in one's head; its feel, as in its "terrible" heat; its physical effect on the human body ("conjunctivitis . . . is common among the nomads"); and its psychological manipulations (the wind tells you that "you are going to go crazy, you are crazy").[47] Despite the aforementioned title of the chapter, the war itself, defending "the bloody fatherland from the moors" is a mere extra to this scene, in which the *īrīfī*, and its maddening voice, are center stage.[48] With the wind coming full circle in its central role propelling the novel, the story closes with Torosantos pointing a gun at his father, who hears the "music of the irifi" as he faces his long-suffering son.[49]

As well as its agentive role, Romeo occasionally uses wind passively as an element to intensify drama. For example, in an early scene that arguably reflects Spain's Franco-era scandal in which the government and Catholic church stole babies from single, Republican, and/or poor mothers in order to give them to pro-Franco families (the mothers were told that their babies had died postpartum), Torosantos's father convinces his wife that her baby, Torosantos's brother, is dead. He later sells the baby, but not before feigning the infant's burial in a shoebox. As Torosantos's mother kneels to God and wails, and her husband digs in the earth, child Torosantos imitates the wind, again and again. The chapter is punctured with speech between Torosantos, who repeatedly utters *uhhhhhhh uhhhhh* in imitation of the wind, and his father, who tells him to "shut up."[50] Torosantos's gathering exhalations in imitation of the wind increase the horror and pathos of the mock burial. Like Siemens's gothic, howling Saharan wind pre-wind-farm-installation, Romeo uses wind to create a vivid, terrifying soundscape of Spanish-Legion-experienced Spanish Sahara. That is, Romeo uses the sounds of the desert winds, as perceived by Spanish soldiers, to evoke the latter's deep-seated fears of the Indigenous. The Legionnaires feel with and through the wind rather than in response to it.

The blood betrayal between father and son in *Discotheque* arguably represents the paternalistic colonizer Spain's betrayal of what it once regarded as its colonial "children," the Saharawis. Likewise, the Spanish-Saharawi siblinghood of Clara and Omar in Reverte's novel can be read as a metaphor for joint destinies, or for transnational solidarity, between Spain and the Saharawis. Family metaphors are also commonly on the tongues of the Spanish solidarity movement with Western Sahara, of which Mayrata is, and Reverte was, a part. I talk of this movement in the singular for ease, but organized Spanish solidarity with Saharawis takes many forms—from the humanitarian, apolitical, and sometimes paternalistic acts of aiding and hosting refugees to direct actions in which activists chain themselves to vessels in protest at the transport of Western Saharan sand to the beaches of the Canary islands. Members of this movement, or these movements, often talk of their "Saharawi brothers and sisters." The sibling metaphor reoccurs in town-*twinning* schemes between Spanish villages and specific Saharawi camps, and also in the *Vacaciones en Paz* scheme, in which Spanish "parents" temporarily "adopt" and house Saharawi children during school summer holidays. I must admit that I, like other solidarity activists and academics with a long-term engagement with the Saharawi people, have my own Saharawi "family" with whom I stay when I visit the Sahara and with whom I maintain bonds of affection.

In the Spanish case, the blood-relative metaphor serves several functions. First, it reminds the Spanish population of its historical relationship with Saharawis and of its state's betrayal of Saharawis through the 1975 Madrid Accords. Second, one should look after one's family and help them when they are in trouble. Betrayals must be atoned for. Blood is thicker than water, *la sangre tira mucho*. The joint effect of these combined functions is a call for solidarity.

Spanish solidarity movements also have their cultural and literary arms. One is the Bubisher project, named after a small desert bird that is said to bring good news. Bubisher, founded in 2008 and run by Spanish and Saharawi writers, began as a mobile library for Saharawi children, which did rounds at the Saharawi refugee camps.[51] Now it has established sedentary as well as several mobile libraries in the camps, has grown into a publisher in its own right, and carries out cultural projects that simultaneously involve Saharawi children in the camps, Spanish children in Spain, Spanish writers, and Saharawi writers. Bubisher's aims are to facilitate access to culture for Saharawi refugee-citizen children, to promote interculturality, and to make visible—internationally—both the Saharawi struggle for independence and Saharawi cultural heritage.[52]

The book *Sand/Water (Arena/Agua, rima/mīa)* is an example of Bubisher's work. It is cowritten by school children from the Wad Salam football club, Smara refugee camp, and Our Lady of the Dove school, Madrid, facilitated by Spanish writer Mónica Rodríguez. The book is bilingual, in Spanish and Arabic. It comprises two parallel stories of long-lost twin sisters, Arena who lives in Spain and Elma who lives in the refugee camps, who magically discover each other's existence. Each of the two stories tells the same tale of discovery and reconciliation, but one is from Arena's perspective and the other from Elma's, representing the sisterly solidarity of Saharawi and Spanish children.

Wind is a vehicle for solidarity in the book. A prologue describes how the children wrote the stories. Mónica Rodríguez facilitated a game in which the Spanish children were given the following prompt: "One day the wind brought a dune to the city," while the Saharawi children were given a similar one: "One day the wind brought a piece of the sea to the desert."[53] The children's plot reveals the backstory of how the two sisters came to be separated: they were lost as babies in a sandstorm. The wind covered Arena in sand

and later uncovered her as some Spanish journalists visiting the Wall of Shame passed by. The journalists adopted Arena and took her to Spain. Meanwhile, Elma was found by her nomadic family and remained in the Sahara. They finally discover their true joint history thanks to the wind that blows a singing and speaking dune to Madrid and a piece of seawater that reflects Arena's face to the camps. Therefore, not only was wind the engine of the prompt that encourages the stories, it also came to shape the plot.

Spanish visual artist Federico Guzman similarly made sand, an aerosol in the desert, a key material for a coproduced animation named after Mohammed Salem's aforementioned poem *The City of Wind*. The animation was inspired by the poem as well as by a conversation Guzman had with another unnamed Saharawi poet. Guzman says that the latter poet told him, "It is impossible to dominate the desert. To survive in it, you have to become the wind."[54] The *City of Wind* animation was made at a sand poetry workshop facilitated by Guzman at San Telmo Museum of Donostia, Spain, in 2016.[55] Guzman led similar workshops in museums and heritage sites in three other Spanish cities that year. All were part of a series of installations named *Tūīza*.

Tūīza is the *Ḥassānīya* word for solidarity or group work of women. The series of installations cocurated by Guzman involved collaborations between himself, other Spanish and Saharawi artists, architects, and poets (most of whom were women), and museum visitors—hence his choice of *Tūīza* as the name for the series. In the case of the sand poetry workshop, the choice of sand as a medium is suggestive of the desert itself as a coproducer. The sand, and indeed all materials for all aspects of *Tūīza*, were brought from the Saharawi refugee camps. All the installations aimed to celebrate cherished aspects of Saharawi identity (chosen by Saharawi coproducers), thereby challenging the status of Saharawis as an "oppressed and forgotten" people.[56] A *ḫaīma*, erected inside each museum or heritage site with *āmlāḥaf* bought from women's collectives in the camps, was the home of all the installations. The sand poetry workshop along with other coproduced artworks were carried out and developed within the *ḫaīma*.

Visitors to the sand poetry workshop, mostly children and their parents of Spanish origin as well as some Saharawi children, animated Mohammed Salem's poem by shifting sand with their fingers over a light box to form representations of Saharawis' traditional nomadic lives in the *bādīa*. The attendees became the wind, drawing shapes in the sand just as the desert wind does. Mohammed Salem and Guzmán provided the soundtrack for the animations—the former reciting her poem and the latter mimicking the sound of the desert winds with his voice.[57]

The cocreated animation is, like Mohammed Salem's original work, imbibed with aeolian aesthetics. Other than hearing its sighs, trickles, whistles, and roars, the main way we know the wind is there is by feeling it. Guzman's tactile medium of finger-driven sand emphasizes Mohammed Salem's own use of synesthesia. The soundtracks of the animations—wind energy in the form of aural poetry and breathed-out sound—showcase Mohammed Salem's alliterative, rhythmic, and assonantal devices, which bring rich aural texture to the original poem and its resultant animation. Likewise, Guzman's use of his own voice mimicking the wind has an aesthetic effect on his audience. Curiously, the sound is almost identical, in terms of pitch and timbre, to that used in the opening of the aforementioned Siemens promotional video. Yet the effect is dazzlingly different. Siemens uses the howling, swooshing sound to suggest desolation, thereby adding further content to its terra nullius narrative. Guzman, however, uses the sound first to emphasize the atmosphere originally created by Mohammed Salem: the sense of wonder and enchantment that one might have felt as a child listening to fairy tales by a fire. Then, in the final frames of the animation, the volume increases and the blown-out sounds are juxtaposed with a flickering flame of sand. The wind's howls and swooshes become, once again, a metaphor for Saharawi resistance. Wind spreads fire, after all, and both are common synecdoches for intifada in Saharawi cultural production.

Guzman's choice of sand as an artistic medium purposefully underlines the metaphorical potential of wind as solidarity. As Guzman explained to me in an interview, Western Sahara's wind is a connecting thread that deposits tangible pieces of Western Sahara, particles of sand, "in Lanzarote, Andalusia, and South America, where it becomes part of the soil in which the rich flora and fauna of the amazon is born."[58] His *Tūīza* put mostly Spanish participants in the place of the desert winds and encouraged them to creatively reflect on Mohammed Salem's ode to Saharawis' cherished nomadic existence in beautiful, nurturing Western Sahara. The installation also challenged the tendency of the international community to forget the Saharawis and thereby show complicity in their oppression. It forged Spanish-Saharawi solidarity in a simultaneously material and imaginary way.

Conclusion

Guzman's coproduction is a tangible example of how the desert winds can inspire solidarity. In Spanish life-writing, wind is used to provoke the reader's empathy with the writer's "lost" time in the Sahara. Likewise, wind is ubiquitously present in the modern Spanish novel on the Sahara. It is rarely a mere part of the background scenery. Named winds—the Sirocco, the *Gatma*, the *irīfi*—whirl around the texts in gusts, setting the scene, indicating the otherwise unwritten emotions of the characters, and telling of their intentions toward others. Spanish desert wind imaginaries can reflect colonial mindsets (Torbado), but they can also offer emancipatory potential (Romeo) and engender solidarity (Reverte, Mayrata, Guzman, and Rodriguez).

This movement of air, although classified and experienced differently, is breathed by Saharawis, Spaniards, and others alike. Discussing the power of international Afro and Arab solidarity with Palestine in the revolutionary 1960s and 70s, Omar Jabary Salamanca argues that solidarity has the power to "subvert artificial divisions and center-periphery paradigms extending far and wide from East to West and South to North."[59] The Spanish texts here join with the Saharawi ones of chapters 5 and 6 in showing how wind imaginaries can unify rather than maintain borders between peoples, calling for solidarity. They have the potential to inform, and be part of, a fairer energy future built on care between humans and more-than-humans across borders.

Conclusion

The Ǧalūāǧ

A *Calima* blows into Andalusia in August 2021. Its fiery, hungry, impatient breath brings, according to some local news sources, the hottest day in the history of Andalusia. There are small tornadoes and a mass of sun-blocking Saharan beige dust. The latter throws us all into a sepia photograph.

The *Calima*'s visit coincides with earthquakes in Granada. We sleep outside in a tent to be on the safe side. The tent is hot. I cannot fathom the heat that Saharawis will be enduring those same days in the Algerian hamada. We go out for the day. My son insists on wearing his *darrāʿa*, a traditional Saharawi sky-blue tunic and trouser set gifted to him by Maimina. We drive to Motril, on Granada's coast. Sand still hangs in the air, creating a midday twilight. Nearing the coast, motorway signage becomes bilingual, Spanish/Arabic. The town itself has several Arab eateries and cybercafés. This is a North African frontier, where two continents are joined by ferry crossings and border controls, moving airs, and subterranean energy cables.

While the boys head to the beach with their father, I visit the port. From here, Siemens Gamesa ships its mills to occupied Western Sahara. Energy systems, like the winds, are transnational. From Teesside to Motril, from Dublin to Nouadibou, so many of us are linked into the oppressive energoregime of Africa's last colony.

Early March 2022. This time, the *Calima* lights up Andalusia with a demonic red glow. The Sierra Nevada resembles Saharan dunes. People laugh with pleasure and wonder as they attempt to ski down the sand. But the dust is pathological, warns Spanish mainstream media: wear masks to prevent respiratory ailments. By March 14, the *Calima*'s burning sighs have spent themselves, although Saharan sand—rendered gold in the southern sun—still blankets Andalusia's window ledges, rooves, and patios. Words in the local newspapers describing Saharan dust recall words used some months earlier: they are words that xenophobically depicted Africans during the Spanish migration "crisis." An invasion. A threat. A cause for fear.

Still March 2022. Prime Minister Pedro Sánchez at last acts over this so-called migration crisis. He means to patch up Spain's diplomatic spat with Morocco over the Ghali incident. In April 2021, Spain allowed Brahim Ghali, leader of Polisario, to be treated for COVID-19 in a Spanish hospital. Morocco responded by recalling its ambassador and opening its shared borders (Spain has two provinces in Africa—Ceuta and Melilla—both bordered on all their inland sides by Morocco) to some ten thousand migrants heading for Europe. Sánchez cannot do with another "invasion" of African immigrants. First, he sacks his foreign minister Arancha González Laya, the fall-woman for the Ghali error. Then, on March 14, 2022, Sánchez pens a letter to Mohammed VI offering support for Morocco's Autonomy Plan, in which Saharawis would remain forever under Moroccan sovereignty, in exchange for the Moroccan King's "co-operation management [*sic*] of migratory flows in the Mediterranean and the Atlantic."[1] This is the first time a Spanish democratic government has formally turned its back on international law and the UN process on the issue of the Western Sahara conflict—another betrayal, forty-seven years after the first.

Najla Mohammed was born in the refugee camps. On her Facebook page on March 17, 2022, she uses windblown sand to reflect on this new Spanish betrayal:

> Tidying the *ḫaīma* while making sure that its doors are shut. It feels like trying to trap a flying balloon. Praying that the *ḫaīma* doesn't fly away and leave us—the family—next to our uncooked food in the corner and our tea tray in the middle, roofless and without walls. Just as if we were sat down for a picnic.
>
> I remember the winter of 2009 very well. We had terrible sandstorms for more than a week.
>
> We shut ourselves in the *ḫaīma* because it's the safest place to be, even though it doesn't really block out the descending sand. We could say it's the best option out of two bad ones.
>
> I remember having short daytime siestas because every 30 minutes or so we had to wake up to dust off all the sand that had formed dunes over our bodies.
>
> We had to push out the sand that kept forming layers, inches think, inside the *ḫaīma*.
>
> Looking outside the *ḫaīma*, each day the scenery was different. The dune moved from one side to another in the patio in front of our *ḫaīma*. It seemed as if the storm was trying to decide on how to decorate our patio with its dunes.

> Sand in our hair, sand down our backs, sand in our food, sand on the tea tray. Sand was a simple fact that no one dared to fight.
>
> At times like that, you simply learn to surrender to the power of nature.
>
> But at the same time you ask yourself: what on earth have our people done to deserve a reality like this?[2]

Ǧalūāğ or *l'ğāğ* is when winds from all four compass rose directions blow at once.[3] Airs from every which way descend on the nomad present in the *l'ğāğ*, and then all those winds blow onward, with a haul of sand, bound for other places, regions, countries. Wind transcends artificial borders, forcing itself on peoples in very different situations in its strange simultaneously material and immaterial way. It carries pieces of the Sahara on transcontinental paths to the Amazon, to the Alps, and to Andalusia. Mohammed's short piece, written in Spanish in the original, forces the (Spanish) reader to reflect on the unequal climate vulnerability of ex-colonizer and re-colonized. Of course, Saharawi nomads knew how to live with the winds that brought sandstorms. But forced sedentarization in the Algerian *ḥammāda* leaves refugees with little possibility for adaptation. Facing the front line of a sandstorm as a refugee in the driest corner of the Sahara is not the same as dealing with the *Calima*'s alleged respiratory irritations at a European ski resort.

The colonial history of Saharan winds stretches back long before pejorative reporting of the *Calima*. Wind made possible the Age of Sail and Europe's colonization of the Americas. It also fueled the hotbed of Euro-American shipwrecks at Cape Boujdour, Western Sahara, the so-called "end of the world." For the most part, mariners' recourse—in their narratives—to hyperbolic heat and "extreme" winds paints a hostile wasteland ripe for a civilizing mission. How we imagine winds tells us how we imagine the places, humans, and other beings of the spaces where those winds blow. The eighteenth- and nineteenth-century mariners see Saharawis as products of their desert weather-world: immune to the imagined pathologies of bright sun and high winds, and all the more barbaric for it. Meanwhile, Saharan sun and winds prove the limits of European colonial prowess in the Age of Sail. The mariners understand that their survival depends on nomads' knowledge of the desert, and they marvel at nomads' ability to read the Saharan sands and winds.

The shipwrecked mariners' windblown narratives, bestsellers in their day and still inspiring new editions, rewrites, and biopics, left their mark on European and American mindsets toward the desert. They formed a foundation for the narratives of twentieth-century Spanish colonizers. Key to colonialism were scientists, researchers, academics. Their publications informed later natural resource exploitation as well as the energy infrastructure that facilitated it. Saharan winds were key to constructing Spanish Sahara as a terra nullius in these publications. The latter also imagine the Saharawis as windblown creatures, whose habits and political structures are likened to the "chaotic" winds and aeolian landforms. The Spanish writers' obsession with the "destructive" winds and the barriers that winds constituted for progressing the colonial mission underline what I call *aeolian anxieties*. Saharan winds exposed the limits of colonial technologies and knowledge as well as the inability of Spaniards to safely travel inland without a nomadic guide. The imaginary and the material shape each other. The Spanish Saharan case shows how starkly this interplay applied to wind imaginaries and energy infrastructure. The Spaniards' deep-seated aeolian anxieties informed the colonial project at all levels, but not least the development, management, and promotion of energy infrastructure in the 1960s, which was presented as a triumph against a "hostile" aeolian environment and which mediated racial segregation.

Just as colonial wind imaginaries shaped Spanish colonial energy infrastructural development, the latter should be understood as an aesthetic arm of colonialism. Domestic canisters, centralized electricity, and fuel tanks changed the aesthetic experiences of Saharawis—the daily smells, tastes, touches, and warmth of Spanish Sahara—exponentially. Reading through the wind forces us to pay attention to senses other than vision. It helps makes clear that energy colonialism is a sensory colonialism as well as a political, economic, and social one.

The legacy of wind-shaped colonial discourses on Western Sahara persists to this day and provides the context for today's energy developments. For the Spanish colonizers, Saharan wind created a wasteland, a terra nullius. Siemens takes this further. Wind itself is waste, terra nullius. In its promotional materials on its wind farms in "Morocco," Siemens draws heavily on sound recordings of the wind and visions of windblown sand. The company claims to have "tamed" the Saharan winds with its mills. Tamed. A word packed with colonial connotations of a civilizing mission, of wild beasts brought to heel, of just victory over the barbaric. Siemens's energy infrastructure is embedded in narratives of economic growth, modernization, and development for a terra nullius, which distorts the reality of natural resource exploitation in a context of protracted conflict, displacement, terrible human rights abuses, and environmental degradation. Of course, it is nothing new for a multinational to draw on colonial clichés to justify dubious activities, but what is of interest is the centrality of wind, and aeolian geomorphology, to such colonial discourses in the case of Siemens. The company relies on aeolian imaginaries to reduce the desert ecosystem, in all its complexity, to wind-as-resource.

Saharawis living in the occupied territory are aware, eyes wide open, that energy infrastructure—its ownership, its management, its reach, the terms of its access, the political and diplomatic work it does—mediates the power of the Moroccan occupation and its corporate partners. The Moroccan occupation enters into, and shapes the possibilities of, daily life in the Saharawi home through (the lack of) electricity cables. Saharawis understand power cuts as a method through which the occupying regime punishes them as a community, fosters ignorance of Moroccan military maneuvers, combats celebrations of Saharawi national identity, enforces a media blockade so that news from Western Sahara does not reach "the outside world," and creates regular dangers in their family homes.

For Saharawis, energy justice is inextricably linked with self-determination. Energy providers, closely linked to the Moroccan state and king, are (seen as) a proxy for the Moroccan occupation in its entirety. Foreign energy developers are despised because they are perceived as agents of colonialism and occupation. The regime attempts to smother those who protest. As I write, Sultana Khaya is still under house arrest, regularly subject to sexual assault whenever police decide to storm her home, and Sidahmed Lemjeyid is now well into his second decade in prison, still tortured often. Colonial wind imaginaries—through which multinational energy developers promote themselves as *taming* hostile, useless, barbaric winds and deserts—breed, underpin, and are part of a colonial (and therefore violent) energoregime.

The Saharan wind imaginaries of European and American mariners, Spanish colonizers, and modern-day energy multinationals overlap considerably. Each builds upon the colonial legacies of those that came before. I concur with colleagues working in energy humanities and social sciences that the hegemonic, Western conception of "energy" is inextricably tied up with colonialism and capitalism because it was born in and of the industrial revolution.[4] Therefore, as several colleagues have convincingly argued, replacing coal and oil with wind, sun, or water as energy sources will not end the political and environmental injustices of the fossil-fuels era unless we rethink energy itself. With this book, I show that colonial wind imaginaries also shape energy developments. These imaginaries have a history partly entwined with the colonial and capitalist phenomena powered by the industrial revolution and therefore also with Western conceptions of energy. But the contribution of wind and wind imaginaries to colonialism also precedes the industrial revolution. A just transition, if that is what we want, therefore requires decolonizing wind imaginaries as much as it does rethinking the Western concept of energy.

Other wind imaginaries exist and might guide us if we let them. Saharawi poetry uses aeolian aesthetics, which are characterized by a decided appeal to the senses through which we know the wind: sound and touch and visions of the windblown. The poetry is informed by wind in terms of its structure, motifs, imagery, and rhetorical devices. It gives readers from outside Western Sahara an insight into nomads' relationship with, ability to read, and dependence on, the Saharan winds, aeolian landforms, and aeolian processes. The numerous named winds and windblown forms are creative, are agents, are useful, have personalities, and facilitate nomadism in numerous ways. They are described with the precise and affectionate words of those whose survival depends on the ability to read aeolian signs. Saharawi texts show wind, and other desert existents, as more-than-human actors to be loved and cared for and as existing in an interconnected, interdependent web. They claim Western Sahara for Indigenous Saharawis not on the basis of colonial, capitalist ownership, but on the basis of love, care, and knowledge. The winds' lessons often come in metaphors. Poets make them messengers, symbols of solidarity—Solidarity between desert existents, but also between Indigenous Saharawis, their cause, and the international community.

What sort of energoregime might such nomadic wind imaginaries inspire? The SADR Energy Ministry's plans for a nomadic energy future in an independent Western Sahara are in chorus with Saharawi wind imaginaries. Existing and envisaged infrastructure in the free zone is small-scale, mindful of more-than-human desert existents, communal and—thinking of nomads—partly portable. It is underpinned by solidarity between Saharawis and other desert beings. For most of their existence, the refugee camps have likewise been powered by small-scale solar and wind installations. In the early days, priority for electrification followed and nurtured Polisario nationalist ideology. For example, in line with Polisario's ideological prioritization of gender equality, the women's training centers were top of the list to receive solar panels.

But forty-seven years of exile wears ideals down. Not everyone can afford a small domestic solar energy system. Refugees would like more comforts in their tents or adobe shelters. As I write, the process of connecting the camps to the Algerian energy grid is underway. This is controversial. The Algerian grid runs on fossil fuels and ties the SADR-in-exile to Algerian infrastructure. But the advantage is that the poorest families allegedly pay nothing for their energy and are subsidized by better-off families. Another controversy is the SADR Mining Authority's assurance agreements with oil companies, which would—in theory—allow the latter to drill in an independent Western Sahara. These agreements contradict the SADR 2021 published plans for exclusively renewable energy in an independent Western Sahara. Some scholars have interpreted the assurance agreements as a ploy to force a legal battle between oil companies in a bid to further Saharawis' right to independence and to challenge the legally questionable oil deals brokered by Morocco in occupied Western Sahara.[5] Whatever the explanation, the SADR's inconsistent energy policies remind us that Indigenous knowledge is situated. It is braided with a certain nomadic way of life in a specific place. It is not possible to simply extract that knowledge and attempt to replicate it in some distant context. While Saharawi wind imaginaries can and have inform(ed) a more just model for renewable energy, there is no nomadic energy future if Saharawis' dispossession and disconnection from their Western Sahara is continued.

Modern Spanish and coauthored Saharawi/Spanish texts offer one more building breeze of hope for our energy futures: solidarity across political borders, beyond one's own community or ecosystem. Modern Spanish writers such as Mayrata or Rodríguez use aeolian metaphors to urge political solidarity with the Saharawi independence cause, while artists such as Gúzman make wind and the windblown their medium in works that encourage a Spanish audience to imagine the Western Saharan desert and its peoples in new ways. What might this aeolian, cross-border solidarity mean in practical terms? Beyond the scope of this book are localized resistance movements up the cable lines, northward. Dunlap and his colleagues have mapped resistance to the social, political, and ecological injustices linked to large-scale wind, solar, and electricity developments all along the cables that link occupied Western Sahara, Morocco, Spain, and France. They conceptualize these injustices as symptomatic of "necropolitical extraction" disguised in green camouflage.[6] Could these resistance movements be linked by (aeolian) solidarity rather than just cables? Could they advocate, together, for small-scale energy systems built on noncolonial imaginaries, respectful of other existents in ecosystems, and mindful of injustices down the supply web if not completely locally produced?

Perhaps one day. Currently though, far from the "supply chain solidarity" advocated by Lennon, we see supply chain betrayal, where the EU stakes part of its "renewable" energy future on the misery of Saharawis.[7] The Spanish government sacrifices their once (or, several legal scholars say, still) colonized subjects in return for Moroccan cooperation in repressing would-be migrants, while Spanish, Italian, and British energy companies amass profits dependent on continued Saharawi oppression.

In the face of criticisms for his policy U-turn, Sánchez repeatedly highlighted that Spain is the biggest supplier of aid to the camps.[8] But Saharawis need political solidarity in their fight for independence, not back-handed "charity." It is the lack of solidarity that has forced the SADR to connect to the Algerian grid, forced Saharawis to embrace return to war with enthusiasm, and forced nomads back from the free zone. Sometimes, when I give talks on the energy systems in occupied Western Sahara, an audience member will lament the price paid by Saharawis but opine that the energy transition is more important than Saharawis' rights in our age of planetary crisis. I would respond, as others have, that "green" extractivism is green only in name.[9] I suggest—I *suggested to* such audience members—that there is (there *was*) an alternative to the oppressive energoregime built by the Moroccan king and foreign corporates. The Saharawi nomadic energy future, based on ethical reciprocity and imaginaries that see winds as creative and skillful and possessing multiple personalities, presents (presented) a far more just model for an energy transition in environmental, cultural, social, political, and economic terms. Since November 2020, I am unsure whether to use present or past tense when discussing this nomadic energy system. Saharawis still aspire to it in a future independent Western Sahara. But the existing wind-powered wells, solar-powered schools and hospitals, and communal kitchen gardens of the liberated territories have, since early 2021, lost most of their nomadic population. Over the last year, the latter have fled Moroccan drone strikes. A week after the start of the new war, Morocco bought $22 million worth of Israeli drones and has since signed a deal to open two Israeli drone factories on its soil, as well as a new strategic partnership with the world's most prolific arms dealer, Lockheed Martin, which includes the opening of an aerospace plant for maintaining and modernizing the Moroccan army's aircraft.[10] How long can the existing nomadic energy system last while this new war rages?

Javier Aroca, in an opinion piece for *El Diario*, reads Sánchez's betrayal through the haze of a Saharan wind. The article, entitled "A *Calima* in Moncloa," states, "Like a new plague, Sánchez's inopportune decision paints Spain in the colour of the Calima, situates it at the mercy of the desert Sirocco and reveals its colonial shame. Other southern winds will come."[11] Meanwhile, on Twitter, several Saharawis and their Spanish supporters likewise invoke the winds in response to the Spanish betrayal. One says, "The calima returns, the sirocco returns, and, as I have said before, we are like thousands of grains of sand that filter into everything, even covering over peoples' ears and shouting . . . Free Sahara!" A Spanish friend in a Sirocco-clad Andalusia tweets, "Now that we abandon the Sahara, the sky seems to tell us that the desert may fall on top of us." With this new stage—this second anti-colonial war—in Western Sahara's history, wind imaginaries of resistance, of resilience, of revenge, and of threat begin to prevail. As the proverbial Saharawi curse goes, "May the Red Wind come after you."

Glossary

āḫīām—plural of *ḫaīma*.
'a'idin—returnees. Saharawis that leave the refugee camps for the occupied territory and receive payments from the Moroccan government for doing so.
āmlāḥaf—plural of *melḥfa*.
asharqiya—an easterly wind.
azafzaf—a wind strong enough to uproot trees.
bādīa—Western Sahara's countryside.
Calima—Spanish-language word for a hot, sand-laden wind.
darrā 'a—Saharawi men's traditional dress.
galb—heart or specific type of Western Saharan rock formation.
galga—an intense sandstorm that restricts visibility.
gallāba—plural of *galb*.
ğalūāğ—a wind that blows from all compass directions at once.
garūāḥ—a peaceful, fresh, and comforting wind.
gatma—dark (black, red, or gray) and blinding sandstorms that appear incredibly rapidly.
gblīya—a southerly wind.
Gdeim Izik—an October 2010 protest camp, razed to the ground a month later by Moroccan authorities, mounted in occupied Western Sahara's *bādīa*. Gdeim Izik also gives its name to the group of political prisoners serving twenty-year to life sentences for their alleged role in violence following the camp's destruction.
ḥaī—neighborhood or quarter (of a town or district).
ḫaīma—traditional Saharawi tent.
ḥammāda—a type of Saharan desert landscape characterized by rocky plateaus.

ḥassānīya—the dialect of Arabic spoken by Saharawis. It is also used in Mauritania and parts of Mali.
īrīfi—hot and powerful easterly wind.
makhzen—the Moroccan ruling class.
melḥfa—Saharawi women's traditional dress.
Polisario Front—the Popular Front for the Liberation of *Saguia el-Hamra* and *Río de Oro*. The Saharawi pro-independence movement founded in 1973 and the only UN-recognized representatives of the Saharawi people.
qabā'el—plural of *qabīla*.
qabīla—tribe.
sāḥlīya—a westerly wind.
Shume—a European word for the *īrīfi* used in the eighteenth and nineteenth centuries.
Simoon—a Spanish word for the *īrīfi*.
tallīya—a northerly wind.
tūīza—solidarity between, or group work of, Saharawi women.
wilāya—province or district.
wilāyāt—plural of *wilāya*.

Notes

Introduction

1. Veale and Endfield, "The Helm Wind," 69.
2. Western Sahara Resource Watch, *Fuelling the Occupation.*
3. Western Sahara Resource Watch, "UK Takes Lead."
4. Pseudonym.
5. Drury, "Global Future and Government Towns."
6. Drury.
7. Personal communication, November 22, 2020.
8. Personal communication, November 19, 2020.
9. Translation by Hamza Lakhal and Joanna Allan. The poem first appeared in Arabic on Facebook, November 25, 2020.
10. Personal communication, November 30, 2020.
11. Salem (pseudonym), interview by Joanna Allan (all interviews are by Joanna Allan unless otherwise noted), occupied El Aaiún, June 27, 2020; Zain (pseudonym), interview, occupied El Aaiún, June 28, 2020.
12. International Court of Justice, "Western Sahara Advisory Opinion."
13. Zunes and Mundy, *Western Sahara*, 114.
14. For details on the various strategies used by Morocco to undermine and block the referendum, see Human Rights Watch, *Keeping It Secret.*
15. Personal communication, November 22, 2020.
16. Drury, "Global Future and Government Towns."
17. Ghayour, *Sirocco.*
18. Ingold, "Eye of the Storm," 104.
19. Flikke, "Matters that Matter, 184.
20. Schmidt, *Hearing Things*, 20.
21. Shalhoub-Kevorkian, "Occupation of the Senses."
22. Hoene, "Sounding City," 364.
23. Ingold, "Earth, Sky, Wind, and Weather," 19.
24. Flikke, "Matters that Matter," 191.
25. Flikke, 192.
26. Klinger, *Rare Earth Frontiers.*
27. Franquesa, *Power Struggles.*
28. Fairhead, Leach, and Scoones, *Greengrabbing.*
29. Hamouchene, "Desertec."
30. Hamouchene.
31. Dunlap and Correa Arce, "'Murderous Energy' in Oaxaca," 49.

32. Dunlap, *Renewing Destruction*, 175.
33. Dunlap, 178.
34. Dunlap, 178.
35. Petrocultures Research Group, *After Oil*.
36. Wilson, Szeman, and Carlson, "On Petrocultures."
37. Black, *Crude Reality*, 8.
38. Howe, *Ecologics*, 191; Boyer, *Energopolitics*, 194.
39. Hughes, *Who Owns the Wind?*
40. Macdonald, "Containing Oil."
41. Boast, *Hydrofictions*.
42. Williams, "This Shining Confluence."
43. *South Atlantic Quarterly* 120, no. 1 (2021).
44. Howe, *Ecologics*.
45. Nye, "Energy Narratives," 25; Daggett, *Birth of Energy*.
46. Lohmann, "White Climate, White Energy."
47. Lohmann, "El cuestionamiento," 4.
48. Lohmann, 1–8.
49. Lohmann, 6.
50. DeBoom, "Climate Necropolitics," 111.
51. Mbembe, "Necropolitics," 11.
52. Ibanga, "Renewable Energy Issues," 120.
53. Ibanga, 123.
54. Asouzu, "Ibuanyidanda (Complementary Reflection)," 19.
55. Bankoff, "Winds of Colonisation."
56. Bankoff.
57. Bankoff.
58. Allan, "Activist Ethics."
59. Ingold, "Footprints through the Weather-World," 132.
60. Kenneally, "Russell Crowe Movie."
61. Pseudonym.
62. White, "Climate, Power, and Possible Futures," 23.
63. See the case of a Siemens blade in a Hull city center piece of public art: White, "Climate, Power, and Possible Futures."
64. Taine-Cheikh, "Ḥassāniyya Arabic," 141.

Chapter 1

1. Azurara, *Discovery and Conquest*.
2. Knight, "Gil Eannes Passes the Point."
3. What Azurara calls *Roses of Saint Maria* are most likely the moss that today is commonly known as Rose of Jericho; Monod, *Les Rosas de Sancta*.
4. Azurara, *Discovery and Conquest*, 31.
5. Lydon, "Writing Trans-Saharan History," 46.
6. Probably a reference to Senegal River, which flowed into the gold-producing Malian empire, although one province of Spanish Sahara was also later named *River of Gold*.
7. Branche, "Inscribing Contact," 9.
8. Azurara, *Discovery and Conquest*, 6–19.
9. Blackmore, *Moorings*.
10. Hughes, *Energy without Conscience*, 144.
11. Hughes, 30.
12. Hughes, 11.
13. Fernández-Armesto, *Before Columbus*, 152.

14. Fernández-Armesto, 161.
15. See several examples in Azurara, *Discovery and Conquest.*
16. Azurara, 95.
17. Azurara, 94. On the use of birds for navigation, Flores Morales detailed in 1949 how Saharawis observed bird flight paths in order to find water sources; Flores Morales, "Tipos y costumbres," 407–8.
18. Cartwright, "Prince Henry the Navigator."
19. Sörlin and Lane, "Historicizing Climate Change," 9.
20. Jackson, *Account of the Empire of Morocco*, 234–35.
21. Baepler, "Barbary Captivity Narrative," 217–18.
22. Thompson, "Introduction," 10.
23. Blackmore, *Manifest Perdition*, xxi.
24. Blackmore, *Moorings*, 52; Thompson, "Introduction," 17.
25. Thompson, "Introduction," 17.
26. Thompson, 17.
27. DeBlieu, *Wind*, 147.
28. Jackson, *Account of the Empire of Morocco*, 226.
29. Jackson, 226.
30. Cochelet quoted in Oldschool, "Dangers of the West Coast of Africa," 1.
31. Riley, *Sufferings in Africa: The Incredible True Story of a Shipwreck*, 43.
32. Sears, *American Slaves and African Masters*, 5.
33. Said, *Orientalism.*
34. Azurara, *Discovery and Conquest*, 27–30.
35. Robbins, *Loss of the Brig Commerce*, 44.
36. Robbins, 44.
37. Saugnier and Brisson, *Voyages to the Coast of Africa*, iv–v.
38. Riley, *Sufferings in Africa: The Incredible True Story of a Shipwreck.*
39. Paddock, *Shipwreck of the Oswego*, chapter 7.
40. Saugnier and Brisson, *Voyages to the Coast of Africa*, 446–500.
41. Black, *British Brig "Surprise" of Glasgow*, 8.
42. Saugnier and Brisson, *Voyages to the Coast of Africa*, 125.
43. Saugnier and Brisson, 125.
44. Cock, *Narrative of Robert Adams*, 123.
45. See, for example, Dean King's use of *Saharawi* in his book *Skeletons on the Zahara.*
46. Snader, "Oriental Captivity Narrative," 275.
47. Riley, *Sufferings in Africa: The Incredible True Story of a Shipwreck*, 42.
48. See chapter 1 of Paddock, *Shipwreck of the Oswego.*
49. Sorenson, "Logging-In."
50. Sorenson.
51. Sorenson.
52. Snader, "Oriental Captivity Narrative," 268.
53. Riley, *Sufferings in Africa: The Incredible True Story of a Shipwreck*, 43.
54. Irving, "Mr. James Irving's Journal," 83.
55. Schwarz, "Introduction," 35.
56. Paddock, *Shipwreck of the Oswego*, 8.
57. Paddock, 12–13.
58. Paddock, 62.
59. Sears, *American Slaves and African Masters.*
60. Cock, *Narrative of Robert Adams.*
61. Scott, "Article X."
62. Cochelet, *Shipwreck of the Sophia*, 26.
63. Saugnier and Brisson, *Voyages to the Coast of Africa*, 359.

64. Paddock, *Shipwreck of the Oswego*, 154.
65. Paddock, 236.
66. Paddock, 284.
67. Paddock, 48.
68. Riley, *Sufferings in Africa: The Incredible True Story of a Shipwreck*, 16.
69. Black, *British Brig "Surprise" of Glasgow*, 25.
70. Riley, *Sufferings in Africa: The Incredible True Story of a Shipwreck*, 119.
71. Paddock, *Shipwreck of the Oswego*, 149.
72. Riley, *Sufferings in Africa: The Incredible True Story of a Shipwreck*, 377.
73. Saugnier and Brisson, *Voyages to the Coast of Africa*, 89–90.
74. Saugnier and Brisson, 35.
75. Paddock, *Shipwreck of the Oswego*, 79.
76. Saugnier and Brisson, *Voyages to the Coast of Africa*, 438.
77. Cock, *Narrative of Robert Adams*, x.
78. Cock, 113.
79. Cock, 33.
80. Cock, xiv.
81. Cock, 139.
82. Cock, xiv–xv.
83. Black, *British Brig "Surprise" of Glasgow*, 22.
84. Riley, *Sufferings in Africa: The Incredible True Story of a Shipwreck*, 481.
85. Paddock, *Shipwreck of the Oswego*, 3.
86. Mahony and Endfield, "Climate and Colonialism."
87. Foxhall, "Interpreting the Tropical Atlantic Climate," 91.
88. Foxhall, 95.
89. Mahony and Endfield, "Climate and Colonialism," 5.
90. Azurara quoted in Blackmore, *Moorings*, 96.
91. Work quoted in Blackmore, 96.
92. Saugnier and Brisson, *Voyages to the Coast of Africa*, 405.
93. Saugnier and Brisson, 467.
94. Haila, "Nature-Culture Dualism."
95. Ponzanesi, *Paradoxes of Postcolonial Culture*.
96. Leyden and Murray, *Discoveries and Travels in Africa*, 281.
97. Sears, *American Slaves and African Masters*, 119.
98. Paddock, *Shipwreck of the Oswego*, 74.
99. Irving, "Mr. James Irving's Journal," 84.
100. Irving, 89; Schwarz, "Introduction," 40.
101. Cock, *Narrative of Robert Adams*, 48.
102. Riley, *Sufferings in Africa, Captain Riley's Narrative*, 132.
103. Paddock, *Shipwreck of the Oswego*, 89.
104. Saugnier and Brisson, *Voyages to the Coast of Africa*, 93.
105. Paddock, *Shipwreck of the Oswego*, 19.
106. See, for example, Cochelet, *Shipwreck of the Sophia*, 40.
107. Cock, *Narrative of Robert Adams*, 19.
108. Black, *British Brig "Surprise" of Glasgow*, 25.
109. Saugnier and Brisson, *Voyages to the Coast of Africa*, 381–82.
110. Saugnier and Brisson, 382.
111. Bandt, "Hearing Australian Identity."
112. Saugnier and Brisson, *Voyages to the Coast of Africa*, 436; Leyden, *Discoveries and Travels in Africa*, 296.
113. Saugnier and Brisson, *Voyages to the Coast of Africa*, 439–40.
114. Riley, *Sufferings in Africa, Captain Riley's Narrative*, 504.

115. Harris, *Weatherland*, 262.
116. Jackson, *Account of the Empire of Morocco*, 283.
117. Jackson, 663.
118. Jackson, 283–84. The expression, transliterated from Arabic, is *āl-baḥr blā mā'*.
119. Cochelet, *Shipwreck of the Sophia*, 34.
120. Riley, *Sufferings in Africa, Captain Riley's Narrative*, 355.
121. Paddock, *Shipwreck of the Oswego*, 86.
122. Robbins, *Loss of the Brig Commerce*, 131.

Chapter 2

1. Ruiz, "Emilio Bonelli."
2. Bárbulo, "Batalla por el fuerte de Villa Cisneros."
3. San Martín, *Western Sahara*.
4. Bárbulo, "Batalla por el fuerte de Villa Cisneros."
5. Galvan Balaguer, *Memoria*.
6. Galvan Balaguer.
7. San Martín, *Western Sahara*, 2.
8. San Martín, 29–30.
9. San Martín, 29.
10. San Martín, 34–35.
11. Rodriguez Esteban, "El mapa del África Occidental Española."
12. Sampedro Vizcaya, "Colonial Politics of Meteorology," 29.
13. Campos Serrano and Trasosmontes, "Recursos naturales," paragraph 6.
14. Díaz de Villegas, *Proyecto de construcciones*, 53–46, 107–48.
15. Communications engineers on behalf of the General Directorate for African Provinces and Posts, *Proyecto de comunicaciones radioeléctricas del Sáhara*, Box 81/11759 Exp. 1.
16. Campos Serrano and Trasosmontes, "Recursos naturales."
17. Camprubí, "Resource Geopolitics," 677.
18. Lalutte, "Sahara," 9.
19. Mariano Alonso Alonso (governor of the Sahara province), letter to Jose Díaz de Villegas y Bustamente (director general of African Posts and Provinces), November 4, 1960. At a time of severe droughts that adversely affected the Saharawi nomads, Mariano Alonso Alonso, then general-governor of the Sahara Province, wrote to the Madrid-based director general of African Provinces and Posts commenting that "due to the drought, there's a serious hunger issue that should not and could not be solved by simply handing out provisions, but by supplying jobs to the Saharawis; you already know that the oil companies need technicians and they don't employ natives" [translation mine]. For complaints by Saharawis of jobs going to Moroccans, see Box 81/11546 of the Africa archive in the Spanish government's General Administration Archive, Alcalá de Henares.
20. Campos Serrano and Trasosmontes, "Recursos naturales."
21. Guinea López, *España y el desierto*, 82.
22. Seth, "Putting Knowledge in Its Place."
23. Mahony and Endfield, "Climate and Colonialism," 6.
24. Sampedro Vizcaya, "Colonial Politics of Meteorology," 28.
25. Hidalgo de Cisneros, "Encuentro con Saint-Exupéry," 121.
26. Hidalgo de Cisneros, 121.
27. Hidalgo de Cisneros, 121.
28. Hernández Pacheco, "El viento en el desierto," 208.
29. Geisler, "New Terra Nullius Narratives," 18.
30. Bedjaoui, *Terra nullius*.
31. Gilbert, *Nomadic Peoples and Human Rights*, 95; Huh, "Title to Territory," 715.

32. Narbón, *Tierra seca*, 7.
33. Watson, *Heaven's Breath*, 247.
34. Narbón, *Tierra seca*, 12–13.
35. Narbón, 13, 23.
36. Narbón, 69.
37. Narbón, 20.
38. Hernández Pacheco, *Sahara español*, 31.
39. Quiroga, "Expedición al Sáhara," 1.
40. Quiroga, 71.
41. Narbón, *Tierra seca*, 13.
42. Congost Tor, "Estudio ornitológico," 196. Regs are desert pavements of closely packed rock. The cause of their formation is debated, but a common theory is that wind gradually removes sands to leave a varnished pavement effect. Wadis are valleys or dry creeks, monadnocks are large rocks or small mountains that rise abruptly from surrounding plains, and *kreb* is a *Ḥassānīya* word for a low, rocky outcropping.
43. Congost Tor, 196.
44. Congost Tor, 202.
45. Hernández Pacheco, *Sahara español*, viii–ix.
46. Livingstone, "Darwinian Hippocratics," 213.
47. Livingstone, 201.
48. Mulero Clemente, *Los territorios españoles del Sahara*, 58.
49. Guinea López, *España y el desierto*, 11.
50. Endfield, "British Women's Emigration Association."
51. Bens, "El sabor de la planta de los pies," 101.
52. Flores Morales, "Tipos y costumbres," 407.
53. Mulero Clemente, *Los territorios españoles del Sahara*, 58.
54. Mulero Clemente, 53.
55. Mulero Clemente, 53.
56. Benítez, "La arena huye del agua," 27.
57. García-Baquero, *Cartografía militar Africana-Española*.
58. Benítez, "La arena huye del agua," 26–27.
59. Saint-Exupéry, *Wind, Sand and Stars*, 48.
60. Cervera, "Expedición al Sáhara."
61. Guinea López, *España y el desierto*, 3.
62. Guinea López, 13.
63. Guinea López, 8.
64. Guinea López, 12.
65. Guinea López, 13.
66. Guinea López, 279.
67. Guinea López, 60.
68. Guinea López, 33.
69. Guinea López, 74.
70. Guinea López, 106.
71. Guinea López, 49.
72. Guinea López, 49.
73. Guinea López, 98.
74. Guinea López, 74.
75. Guinea López, 74.
76. Guinea López, 161.
77. D'Almonte, "El Sahara español," 63. *Gallaba* is the plural of *Galb*, which is *Ḥassānīya* for both "heart" and "large rock or mountain."
78. Sampedro Vizcaya, "Colonial Politics of Meteorology," 45.

79. Cervera, "Expedición al Sáhara," 56.
80. Rodriguez Esteban, "El Mapa del África Occidental Española," paragraph 36.
81. Hidalgo de Cisneros, "Encuentro con Saint-Exupéry," 122.
82. Congost Tor, "Estudio ornitológico," 195.
83. Narbón, *Tierra seca*.
84. Guinea López, *España y el desierto*, 131.
85. Guinea López, 53.
86. Rodriguez Esteban, "El Mapa del África Occidental Española."
87. Rodriguez Esteban.
88. Rodriguez Esteban, paragraph 37.
89. Domenech Lafuente, *Algo sobre Río de Oro*. See also Rodriguez Esteban, "El Mapa del África Occidental Española," footnote 14.
90. Flores Morales cited in Rodriguez Esteban, "El Mapa del África Occidental Española," paragraph 45.
91. Mahmoud Alina, interview by Hamza Lakhal, occupied El Aaiún, May 27, 2019.
92. Mahmoud Alina, interview by Hamza Lakhal, occupied El Aaiún, May 27, 2019.
93. Moiti Mohammed, personal communication, March 29, 2023.
94. Moiti Mohammed, personal communication, March 29, 2023.
95. Zain (pseudonym), interview by Mahmoud Lemaadel, occupied El Aaiún, June 28, 2020.
96. San Martín, *Western Sahara*, 49–50.
97. Isidoros, "Silencing of Unifying Tribes," 173–74.
98. Fernando Sanjurjo de Carricarte, letter to the technical director of the Díaz de Villegas Power Station, November 8, 1967. See, for example, complaints from El Aaiún's swimming pool, which was unable to run its motors or push water through its filters.
99. Governor of the Sahara Province, letters to the director general of African Posts and Provinces, October 18, 1965, and July 23, 1966. For example, the governor-general writes to Madrid of the "constant pressures" on him from "native and European residents" in his requests for a power station in Smara.
100. Governor of the Sahara Province, letter to the director general of African Posts and Provinces, March 20, 1967; governor of the Sahara Province, letter to the director general of African Posts and Provinces, July 23, 1966; Angel Enriquez Larrondo from the Public Works office, letter to Jose Díaz de Villegas y Bustamante, director general of Posts and African Provinces, El Aaiún, October 15, 1966; Fernando Sanjurjo de Carricarte, letter to the technical director of the Diaz de Villegas power station, El Aaiún, November 8, 1967; head of Public Works office, letter to the secretary general, El Aaiún, November 26, 1968; secretary general, letter to the head engineer of the Public Works office, El Aaiún, May 26, 1970; Fernando de Sandoval y Coig, letter to Antonio Sánchez del Toro Molino (commander construction engineer), March 16, 1971. For more on the dangerousness of high-tension power lines, see governor of the Sahara Province, letter to the director general of African Posts and Provinces, March 20, 1967.
101. Governor of the Sahara Province, letter to the director general of African Posts and Provinces, July 23, 1966.
102. Chief engineer of public works, letter to the secretary-general of the Sahara Province, November 26, 1968.
103. Secretary-general of the Sahara Province, letter to the chief engineer of public works, May 26, 1970.
104. Fernando de Sandoval y Coig, letter to Antonio Sánchez del Toro Molino (commander construction engineer), March 16, 1971.
105. Mayor of El Aaiún, letter to the secretary-general of the Sahara Province, October 23, 1971.
106. Angel Enriquez Larrondo (Public Works), letter to the director general of African Posts and Provinces, October 15, 1966.

107. Governor of the Sahara Province, letter to the director general of African Posts and Provinces, July 23, 1966.
108. Governor of the Sahara Province, letter to the director general of African Posts and Provinces, July 23, 1966; general governor of the Sahara Province, letter to the director general of African Posts and Provinces, April 27, 1969; Government Delegate, *Permission Document*; Engineer of Pathways, Canals and Ports, General Government of the Sahara Province, Provincial Service for Public Works, *Extension of the Smara Power Station*.
109. Zrug (pseudonym), interview by Mahmoud Lemaadel, June 28, 2020.
110. Mahjoub (pseudonym), interview by Mahmoud Lemaadel, September 9, 2020.
111. Saharawis, *Petition of Signatures*.
112. Governor of the Sahara Province, letter to the director general of African Posts and Provinces, October 18, 1965.
113. See, for example, Harrison, "American South."
114. Technical director of Díaz de Villegas power station, "Solicitud viviendas del I.N.V." See, for examples, the case of Brahim Uld Moh Uld Lahadih, an electrician at La Guera power station, who lobbied for a higher salary, and of Mohammed Mohtar uld Abdelahe and Abdel-Krim Azman Matl-la, who requested lodgings along with their Spanish colleagues while working at El Aaiún's power station.
115. Kirshner, Castán Broto, and Baptista, "Energy Landscapes in Mozambique," 1055.
116. Shalhoub-Kevorkian, "Occupation of the Senses."
117. Atlas S.A. combustibles y lubrificantes, *Dossier*, n.p.
118. Atlas S.A. combustibles y lubrificantes.
119. Atlas S.A. combustibles y lubrificantes.
120. Atlas S.A. combustibles y lubrificantes.
121. Atlas S.A. combustibles y lubrificantes.
122. Atlas S.A. combustibles y lubrificantes.
123. Nguia Hassan (pseudonym), interview by Mahmoud Lemaadel, occupied El Aaiún, October 19, 2019.
124. Schivelbusch, *Disenchanted Night*, 28–29.
125. Schivelbusch.
126. Mrázek, *Engineers of Happy Land*.
127. See, for example, Caffarena Aceña, *Informe sobre los daños causados*.
128. See Box S.2.179, top. 53/46, 107–48, 103, Exp. 1 in the Spanish Government's General Administrative Archives.
129. For more on this, see Anderson, "Weather Ship."
130. Alvarez de Benito, *File of Research*.
131. See Box S.2.116, top. 65/76, 401–78.709, Exp. 2 in the Spanish Government's General Administrative Archives.
132. Mahmoud Lemaadel, phone conversation, February 15, 2020.
133. Malainin Lakhal, phone conversation, March 17, 2021.
134. Nguia Hassan, interview by Mahmoud Lemaadel, occupied El Aaiún, October 19, 2020.
135. Mahmoud Alina, interview by Hamza Lakhal, occupied El Aaiún, May 27, 2019.
136. Haidar, "La juventad de 'Irifi,'" 272.
137. Haidar, 284.
138. For more on the emergence of Saharawi nationalism in the 1960s and 1970s, see San Martín, *Western Sahara*; and Baba Miské, *Front Polisario*.
139. San Martín, *Western Sahara*, 71.
140. For more on this war, see San Martín, *Western Sahara*, 66–73; Drury, "Anticolonial Irredentism"; and Correale and López Bargados, "Rashōmon au Sahara occidental."
141. San Martín, *Western Sahara*, 66–73.
142. Bashir Mustapha Sayed, cited in San Martín, 81.

143. For more on this incident, see Bárbulo, *La historia prohibida*, 136–41; Diego Aguirre, *Historia del Sahara español*; and Martínez Milán, "Empresa pública y minería," 920–21.
144. Bonelli, *El Sahara*, 228–29.

Chapter 3

1. Pseudonym.
2. For a review of legal studies publications on natural resource exploitation in occupied Western Sahara, see the relevant section of Allan and Ojeda-García, "Natural Resource Exploitation." See also Western Sahara Resource Watch, "Court Annuls EU Deals."
3. General Court of the European Union, "The General Court Annuls the Council Decisions."
4. Pseudonym.
5. Pseudonym.
6. Africa Intelligence, "Dakhla Tourism Empire."
7. Pseudonym.
8. Pseudonym.
9. Pseudonym.
10. Merriman, "Nonviolent Action."
11. Stephan and Mundy, "A Battlefield Transformed," 21–22.
12. Allan, "Activist Ethics," 99.
13. Hopman, "Covert Qualitative Research," 3.
14. Hopman, 15.
15. Human Rights Watch, "Morocco/Western Sahara."
16. Allan, "Activist Ethics," 100.
17. For more on the US organization, see their web page: https://justvisitwesternsahara.org/our-work/. For more on Sharp's theorization of nonviolent struggle and the role of solidarity, see Sharp, *How Nonviolent Struggle Works*.
18. Saleh quoted in Allan, "Activist Ethics," 95.
19. Translation by Hamza Lakhal. For more on this song and the poem from which its lyrics is drawn, see Ruano Posada, "The Sahara Is Not for Sale: The Song of a Revolution."
20. Pseudonym.
21. *Irish Times*, "Fortress Europe or Fair Trade." A €148 million EU investment in Morocco's border controls on the northern frontier has forced migrants to take the even more perilous Atlantic route between Western Sahara and the Canary Islands. Seventeen thousand people arrived in the latter during the first ten months of 2021. See Greene and Gilmartin, "Europe's Migrant Crisis."
22. Long-term bonds of affection between foreign solidarity activists' families and Saharawi families are further discussed in chapter 8.
23. Western Sahara Resource Watch, *Dirty Green March*, 13.
24. Western Sahara Resource Watch, 13.
25. Zain (pseudonym), interview by Mahmoud Lemaadel, El Aaiún, June 28, 2020.
26. Western Sahara Resource Watch, "Saharawi Family Home Lost."
27. See especially and Luque-Ayala and Silver, *Energy, Power, and Protest*.
28. Angel, "New Municipalism and the State."
29. Kirshner, Castán Broto, and Baptista, "Energy Landscapes in Mozambique."
30. Enns and Bersaglio, "'New' Mega-Infrastructure Projects."
31. Power et al., "Political Economy of Energy Transitions."
32. Dunlap, *Renewing Destruction*; Howe and Boyer, "Aeolian Politics"; Jabary Salamanca, "Unplug and Play."
33. Lloyd and Wolfe, "Settler Colonial Logics."
34. Western Sahara Resource Watch, "Hunger Striking against EU Fisheries."

35. See video by Western Sahara Resource Watch, "Hunger Striking against EU Fisheries."
36. Western Sahara Resource Watch, *Totally Wrong*.
37. Lucente, "Emirati Firm to Build."
38. Jabary Salamanca, "Unplug and Play."
39. Khatib, *Morocco Country Profile*, 4.
40. Eau du Maroc, "Réseau electrique au Sahara."
41. ONEE, *ONEE au Maroc*.
42. Boulakhbar et al., "Towards a Large-Scale Integration," 8; Grundy, "Octopus Invests."
43. Boulakhbar et al., "Towards a Large-Scale Integration," 8.
44. North Africa Post, "Morocco Reaps Diplomatic Gains"; Bennis, "Morocco's Contemporary Diplomacy."
45. Ngounou, "Morocco."
46. Hassan, "Morocco Flirts with Ethiopia."
47. *North Africa Post*, "Morocco Reaps Diplomatic Gains."
48. ONEE, *ONEE au Maroc*, 2–3.
49. ONEE.
50. Goldthau, "Energy Diplomacy."
51. Fernández Camporro, "The King's Speech."
52. Nye, *Hard, Soft and Smart Power*.
53. Fairhead, Leach, and Scoones, *Greengrabbing*; Dunlap, *Renewing Destruction*.
54. Western Sahara Resource Watch, *Greenwashing Occupation*.
55. Western Sahara Resource Watch, 10.
56. OCP Group, *2019 Sustainability Report*.
57. Raminder Kaur, "Southern Spectrums," 23.
58. Power et al., "Political Economy of Energy Transitions."
59. Africa Intelligence, "Masen to Build Noor Solar Farm."
60. Moustakbal, *Moroccan Energy Sector*, 4.
61. Hagen and Pfeifer, *Profit over Peace in Western Sahara*, 12.
62. Kohn and Reddy, "Colonialism."
63. Steinman, "Decolonization Not Inclusion," 221.
64. See, for example, Barker, "Locating Settler Colonialism"; Tuck and Yang, "Decolonization Is Not a Metaphor"; Veracini, "Settler Colonialism and Decolonisation"; Veracini, *Settler Colonialism*; Veracini, "Introducing: Settler Colonial Studies"; and Wolfe, *Settler Colonialism and the Transformation of Anthropology*.
65. Veracini, "Settler Colonialism and Decolonisation," 1.
66. Wolfe, *Settler Colonialism and the Transformation of Anthropology*, 2.
67. Veracini, "Introducing: Settler Colonial Studies."
68. Veracini, 2.
69. Veracini, "Settler Colonialism and Decolonisation," 6.
70. Veracini, "Introducing: Settler Colonial Studies," 3.
71. See chapter 5 of Allan, *Silenced Resistance*.
72. Human Rights Clinic of Cornell Law School and Clinic on Fundamental Rights of the Université de Caen Basse-Normandie, *Report on the Kingdom of Morocco's Violations*, 13.
73. See, for example, Isidoros, *Nomads and Nation-Building*; Pablo San Martín, *Western Sahara*, 10; and Porges and Leuprecht, "Puzzle of Nonviolence."
74. Athreya, "Perestroika and the Third World," 28–29.
75. I borrow "environmental violence" from Sakshi Aravind, who uses it to describe the phenomena of large corporations destroying indigenous lands. Aravind, "Many Entanglements," 73.
76. Saharawi Arab Democratic Republic, *First Indicative Nationally Determined Contribution*, 19. See also Together to Remove the Wall, "Impacts of the Wall."
77. Saharawi Arab Democratic Republic, *First Indicative Nationally Determined Contribution*, 19.

78. Nixon, *Slow Violence*, 2.
79. Crook and Short, "Genocide–Ecocide," 167.
80. Short, *Redefining Genocide*, 6.
81. Volpato and Howard, "Material and Cultural Recovery of Camels."
82. Weexsteeny, "Fighters in the Desert."
83. San Martín and Allan, *Largest Prison in the World*; Saharawi Arab Democratic Republic, *First Indicative Nationally Determined Contribution*, 19.
84. Shaheed, *Report of the Independent Expert*, 18.
85. Human Rights Clinic of Cornell Law School and Clinic on Fundamental Rights of the Université de Caen Basse-Normandie, *Report on the Kingdom of Morocco's Violations*, 24.
86. Capitalist land grabs and land privatization policies are negatively impacting on mobile pastoralists elsewhere. See, in particular, Bassett, "Mobile Pastoralism."
87. Association for the Monitoring of the Resources and the Protection of the Environment (AMRPENWS), "Environmental Issues in Western Sahara."
88. Western Sahara Resource Watch, *Label and Liability*.
89. Yang et al., "Environmental Impacts"; Saharawi Natural Resource Watch, *Saharawis*.
90. Amoko, "The 'Missionary Position.'"
91. Amoko, 312.
92. Laclau and Mouffe, *Hegemony and Socialist Strategy*.
93. Howarth, *Discourse*, 3.
94. Wolfe, *Settler Colonialism and the Transformation of Anthropology*, 2.
95. Siemens, "About Us."
96. Dachverband der Kritischen Aktionärinnen and Aktionäre, *Counterproposals*.
97. The Siemens press release is quoted in full in WSRW, "Siemens with Press Release."
98. See the Environmental Justice Atlas for more on this press release. Environmental Justice Atlas, "Wind Power Plants."
99. "Greenwashing."
100. See WSRW, "Siemens with Press Release."
101. Crooks, and McGee, "GE and Siemens."
102. OCP Group, *2019 Sustainability Report*, 119.
103. Moore, "Introduction."
104. Destruction of Western Sahara's fisheries is another environmental issue. Historically, the waters of Western Sahara were among the world's richest fishing stocks, but according to Greenpeace, catches are now in decline. Greenpeace, *Exporting Exploitation*, 19.
105. Greenpeace, *Exporting Exploitation*; Saharawi Natural Resource Watch, *Saharawis*.
106. Oceanic Dévelopement, *Evaluation ex-post du Protocole*.
107. FAO quoted in Greenpeace, *Exporting Exploitation*, 19.
108. Western Sahara Resource Watch, "Check Out Your Fish Here"; Western Sahara Resource Watch, "Morocco Continues to Discard By-Catches"; Western Sahara Resource Watch, "Trawler in Occupied Western Sahara."
109. Western Sahara Resource Watch, "Self-Immolation by Moroccan Sea Captain."
110. Jabary Salamanca, *When Settler Colonialism Becomes 'Development,'* 1–3.
111. Jabary Salamanca.
112. Apter, *Rethinking Development*; Mcfarlane and Rutherford, "Political Infrastructures."
113. Mcfarlane and Rutherford, "Political Infrastructures," 364.
114. Rankin, "Infrastructure and the International Governance," 63.
115. Zanon, "Infrastructure Network Development."
116. Kooy and Bakker, "Technologies of Government."
117. Harris, "How Did Colonialism Dispossess?"
118. Harris.
119. Said, *Orientalism*.
120. Siemens, "Supporting Morocco's Sustainable Development."

121. Volpato, "Material and Cultural Recovery of Camels."
122. Volpato.
123. Volpato.
124. See, for example, Siemens, *Pictures of the Future*, 33.
125. Franquesa, *Power Struggles*, 230.
126. Siemens, "Turning the Wind's Energy."
127. Hunold and Leitner, "'Hasta La Vista, Baby!'"
128. Hunold and Leitner, 693.
129. Beck, "Without Form and Void."
130. Siemens, "Turning the Wind's Energy."
131. The video has been archived and is available here: http://players.brightcove.net/1813624294001/SJYnxX9G_default/index.html?videoId=5276877973001.
132. Hamouchene, "Desertec."
133. Hamouchene.
134. Van Son, "Desertec Industries."
135. Boletsi, *Barbarism and Its Discontents*.
136. Emphasis in original. Dall'omo, "We Are All Empowered."
137. Siemens, *Pictures of the Future*.

Chapter 4

1. Pseudonym.
2. Mahjoub, interview by Mahmoud Lemaadel, occupied El Aaiún, September 9, 2020.
3. For more on corruption, environmental devastation, and discrimination in the fishing industry of occupied Western Sahara, see Observatorio de derechos humanos y empresas en el Mediterráneo, *Los tentáculos de la ocupación*.
4. Mahmoud, interview by Hamza Lakhal, occupied El Aaiún, May 17, 2020.
5. Mahmoud, interview by Hamza Lakhal, occupied El Aaiún, May 17, 2020.
6. Barona Castañeda, "Memorias de una resistencia."
7. Allan and Lakhal, *Acting with Impunity*; Allan, *Silenced Resistance*; Martín Beristain and González Hidalgo, *El oasis de la memoria*.
8. The renowned NGO Freedom House gave Western Sahara a score of 4 out of 100, putting it one place behind North Korea, in its 2022 global freedom ratings. See Freedom House, "Freedom in the World 2022."
9. Brew-Hammond, "Energy Resources in Africa."
10. Winther and Wilhite, "Tentacles of Modernity."
11. Power et al., "Political Economy of Energy Transitions," 504.
12. Lemanski, "Infrastructural Citizenship."
13. Baptista, "Maputo."
14. Enns and Bersaglio, "'New' Mega-Infrastructure Projects."
15. Power et al., "Political Economy of Energy Transitions."
16. Dunlap, *Renewing Destruction*; Howe and Boyer, "Aeolian Politics"; Jabary Salamanca, "Unplug and Play."
17. Boyer, "Energopower," 197.
18. Rogers, "Energopolitical Russia."
19. Mitchell, *Carbon Democracy*.
20. Foucault, *History of Sexuality*.
21. Mitchell, *Carbon Democracy*.
22. Palmer et al., "Oppression and Power."
23. Cudd, "How to Explain Oppression," 21.
24. Márquez, *Non-Democratic Politics*, 234.
25. Zafirovski, *Protestant Ethic and the Spirit of Authoritarianism*, 137.

26. Pseudonym.
27. All first names indicate pseudonyms for interviewees.
28. Pseudonym.
29. Enloe has made this observation with regard to the situation of women living in Iraq during the US invasion (2017). For more on the gendered implications of infrastructural failures, see also Lawhon et al.,"Heterogeneous Infrastructure Configurations."
30. Pseudonym.
31. Fadel, interview by Hamza Lakhal, occupied El Aaiún, May 28, 2019.
32. Hartan (pseudonym), interview by Mahmoud Lemaadel, occupied El Aaiún, October 19, 2020.
33. Chikowero, "Subalternating Currents"; Winther, *The Impact of Electricity*.
34. Jabary Salamanca, "Unplug and Play."
35. Harrison, "American South."
36. Baptista, "Maputo," 121.
37. Nguia (pseudonym), interview by Mahmoud Lemaadel, occupied El Aaiún, October 19, 2020.
38. Dadi (pseudonym), interview by Mahmoud Lemaadel, occupied El Aaiún, October 19, 2020.
39. Baptista, "Maputo."
40. Zrug, interview by Hamza Lakhal, occupied El Aaiún, June 28, 2020.
41. Salka (pseudonym), interview by Mahmoud Lemaadel, occupied El Aaiún, September 9, 2020.
42. Omar (pseudonym), interview by Mahmoud Lemaadel, occupied El Aaiún, September 8, 2020.
43. Mahmoud, interview by Hamza Lakhal, occupied El Aaiún, May 27, 2019.
44. Najem (pseudonym), interview by Mahmoud Lemaadel, occupied El Aaiún, September 7, 2020.
45. Verdeil, "Beirut, Metropolis of Darkness," 163–65.
46. Silver, "Incremental Infrastructures," 795.
47. Verdeil, "Beirut, Metropolis of Darkness."
48. Zrug, interview by Hamza Lakhal, occupied El Aaiún, June 28, 2020.
49. Lemanski, "Infrastructural Citizenship."
50. Chapman et al., "Conversation 2"; Rupp, "Considering Energy."
51. Silver, "Incremental Infrastructures."
52. Zrug, interview by Hamza Lakhal, occupied El Aaiún, June 28, 2020.
53. Zrug, interview.
54. Zrug, interview.
55. Zrug, interview.
56. Nguia, interview by Mahmoud Lemaadel, occupied El Aaiún, October 19, 2020.
57. Observatorio de derechos humanos y empresas en el Mediterráneo, *Los tentáculos de la ocupación*.
58. Nguia, interview by Mahmoud Lemaadel, occupied El Aaiún, October 19, 2020.
59. Najem, interview by Mahmoud Lemaadel, occupied El Aaiún, September 7, 2020.
60. Hartan, interview by Mahmoud Lemaadel, occupied El Aaiún, October 19, 2020. Kosmos Energy was among a partnership of three energy companies that were the first to drill for oil off Western Sahara's coast. Likewise, San Leon was part of the partnership that was the first to drill onshore.
61. Dadi, interview by Mahmoud Lemaadel, occupied El Aaiún, October 19, 2020.
62. Ahmed, interview by Mahmoud Lemaadel, occupied El Aaiún, October 19, 2020.
63. Salka (pseudonym), interview by Mahmoud Lemaadel, occupied El Aaiún, September 9, 2020.
64. Mahjoub, interview by Mahmoud Lemaadel, occupied El Aaiún, September 9, 2020.

65. Fadel, interview by Hamza Lakhal, occupied El Aaiún, May 28, 2019.
66. Mahmoud, interview by Hamza Lakhal, occupied El Aaiún, May 17, 2020.
67. Zain (pseudonym), interview by Hamza Lakhal, occupied El Aaiún, June 28, 2020.
68. Western Sahara Resource Watch, *Dirty Green March*; Western Sahara Resource Watch, "Saharawi Family Home Lost."
69. Salem (pseudonym), interview by Hamza Lakhal, occupied El Aaiún, June 27, 2020.
70. Baptista, "Maputo."
71. DeBoom, "Climate Necropolitics"; Mbembe, "Necropolitics."
72. Amnesty International, "Urgent Action: Sahrawi Activist Raped."
73. Amnesty International, "Morocco: Sahrawi Activist Raped."
74. Frontline Defenders, "Sultana Khaya was Sexually Assaulted."
75. Frontline Defenders, "House Arrest and Physical Assault."
76. Sultana Khaya, video interview by Joanna Allan, January 19, 2022.
77. Dunlap, "Does Renewable Energy Exist?," 96.
78. Sultana Khaya, video interview by Joanna Allan, January 19, 2022.
79. Sultana Khaya, video interview.
80. Allan, *Silenced Resistance*.
81. Sørfonn Moe, *Observer Report*, 258.
82. The report by Associação de Cooperação e Solidariedade Entre os Povos (ACOSOP) is published in Sørfonn Moe, 250–65.
83. Corell, *Letter Dated 29 January 2002*.
84. ACOSOP report published in Sørfonn Moe, *Observer Report*, 250–65.
85. Sørfonn Moe, 258.
86. Sørfonn Moe, 34.
87. Pseudonym.
88. Pseudonym.
89. Pseudonym.

Chapter 5

1. Allan, "The Saharawi 'Friendship Generation'"; Allan, "Colonialism and Patriarchy"; Allan, "Nationalism, Resistance and Patriarchy." Parts of this chapter have been modified from an article that originally appeared in the *Bulletin of Hispanic Studies*: Allan, "Decolonizing Renewable Energy."
2. Howe, "Aeolian Politics," 31.
3. Howe, *Ecologics*; Williams, "This Shining Confluence"; Szeman, "Conjectures on World Energy Literature," 278. Emerging work on solar imaginaries is also of interest. See, for example, Szeman and Barney, "From Solar to Solarity."
4. Szeman et al., *On the Energy Humanities*, 3.
5. Macdonald, "Containing Oil," 291.
6. Macdonald, "Oil and World Literature."
7. Niblett, "Oil on Sugar."
8. The poem was first published in Boisha's second collection of poetry *Ritos de jaima*.
9. Tuan, "Space and Place." As Yi-Fu Tuan argues, human senses of place are not only imagined and the result of symbolic projection but also formed by long sensory associations with the environment.
10. Boisha, *Ritos de jaima*.
11. Ellison, "La amada Tiris."
12. Volpato, "Material and Cultural Recovery of Camels."
13. Ellison, "La amada Tiris," 73.
14. San Martín and Bollig, *Treinta y uno*. Translation by Ed Atkins, printed in 2007.
15. On traditional Saharawi poetry, see Awah, "Literatura oral y transmisión."

16. Limam Boisha, interview by Joanna Allan, Madrid, July 6, 2018.
17. Volpato, "Material and Cultural Recovery of Camels."
18. Volpato and Kumar Puri, "Dormancy and Revitalization," 206.
19. Volpato and Kumar Puri.
20. See, for example, Yuval-Davis, *Gender and Nation.*
21. McClintock, *Imperial Leather.*
22. Howe, *Ecologics*, 28, 34.
23. Poem first published in Mohammed Salem, *Nada es eterno*. Available online at http://literaturasaharaui.blogspot.com/2010/04/poemas-saharauis-para-crecer-nada-es.html.
24. Literally hundreds of UN Security Council resolutions demand a free and fair self-determination referendum for the Saharawi people, and UN General Assembly resolutions denounce the "continued occupation of Western Sahara by Morocco." See, for example, United Nations General Assembly, *Resolution 35/19.*
25. Corell, *Letter Dated 29 January 2002.*
26. Mohammed Salam, *Nada es eterno.*
27. Awah, "Literatura oral y transmisión."
28. This is also the case for the Indigenous people of the Namib desert. See BBC World Service, "The Sounds of the Namib Desert."
29. Ingold, "Earth, Sky, Wind, and Weather."
30. Mohammed Salem nods to this ignorance through the repeated lines "veo lo que nadie ve, / siento lo que nadie siente" ["I see what no one sees, / I feel what no one feels"] as if to recognize that her Western audience is unable to appreciate the desert attributes that she sees and feels so clearly.
31. For more on reading weather patterns in dunes and ripples, see Fenton, "Sand Waves in the Desert."
32. Taine-Cheikh, "Following the Tracks of the Moors," 203.
33. Fatma Galia Mohammed Salem, telephone interview by Joanna Allan, September 18, 2018.
34. Odartey-Wellington, "Walls, Border, and Fences," 5.
35. Odartey-Wellington, 5.
36. Szeman and Boyer, "Introduction," 3.

Chapter 6

1. Al Montassir, *Galb'Echaouf*, 5:18.
2. Al Montassir, *Al Amakine.*
3. Al Montassir, *Galb'Echaouf*, 9:12.
4. Al Montassir, 9:29.
5. Al Montassir, 9:34.
6. Al Montassir, 12:30.
7. Al Montassir, 14:15.
8. Al Montassir, 15:13.
9. Abdessamad Al Montassir, personal communication, July 8, 2021.
10. Merleau-Ponty, *Phenomenology of Perception.*
11. Al Montassir speaking at Bodies of Occupation event, Nottingham University, October 22, 2021.
12. Al Montassir speaking at Bodies of Occupation event.
13. On colonial discourse, see, for example, Quijano, "Coloniality of Power."
14. Al Montassir, Dossier Artistique.
15. Abdessamad Al Montassir, personal communication, June 11, 2021.
16. Al Montassir, *Galb'Echaouf*, 10:18.
17. Abdessamad Al Montassir, personal communication, June 11, 2021.

18. This is my own reading of the poem, but it is based on conversations with Zaim Allal (Aaiún camp, February 2023) and personal conversations with Hamza Lakhal.
19. The poem is reproduced here with the permission of its author, Zaim Allal. It is translated by Hamza Lakhal from a *Ḥassānīya* version transcribed by Al Montassir and then extended by Zaim Allal (the extra verse was transcribed and translated by Hamza Lakhal).
20. Boulay, "'Returnees' and Political Poetry." For an extended discussion on returnees and their political subjectivities, see Drury, "Disorderly Histories."
21. See Observatorio de derechos humanos y empresas en el Mediterráneo, *Los tentáculos de la ocupación*.
22. Western Sahara Resource Watch, *Fuelling the Occupation*, 3.
23. Berkson, "Voices of a Lost Homeland."
24. For the Taiwanese example, see Kuo, "From Solidarity to Antagonism." For the cases of English, French, German, and Spanish, see Brown and Gilman, "Pronouns of Power and Solidarity."
25. Van Gelder, "The Abstracted Self," 29.
26. Ellison, "La amada Tiris." This article highlights how common the second person voice is in modern Saharawi poetry in Spanish.
27. Lakhal, *Destiny in Windblown Poetry*.
28. For more on repression suffered by Hamza due to his work, see Lakhal, "Fighting for a University."
29. Translation mine. Original poem in Arabic published in Lakhal, *Destiny in Windblown Poetry*, 30–36.
30. Criado-Hernández, Moreno-Medina, and Dorta-Antequera, "Otra imagen del desierto," 10.
31. Hamza Lakhal, personal communication, March 13, 2022.
32. Hamza Lakhal, personal communication, March 14, 2022.
33. Davila, "East Winds and Full Moons," 17. In classical Arabic love poetry (ghazal), the east wind was commonly bearer of news of the beloved.
34. This poem was published on Facebook in April 2021.
35. Pseudonym.
36. Pseudonym.
37. Pseudonym.
38. Zaim Allal, interview by Joanna Allan, Saharawi refugee camps, October 10, 2019.
39. Zaim Allal, interview.
40. Zaim Allal quoted in Ruano Posada and Solana Moreno, "Strategy of Style," 40.
41. Allal quoted in Ruano Posada and Solana Moreno, 40.
42. Zaim Allal, interview by Joanna Allan, Saharawi refugee camps, October 10, 2019.
43. Zaim Allal, interview.
44. Bruchac, "Indigenous Knowledge and Traditional Knowledge," 3814.
45. Nguia Hassan, interview by Mahmoud Lemaadel, El Aaiún, October 19, 2020.
46. Future Learn, "Living Ecosystems."
47. Future Learn, "Living Ecosystems."
48. Abram, *Spell of the Sensuous*, 106.
49. Belfrage cited in Bandt, "Hearing Australian Identity."
50. Awah, *Tiris*.
51. Gimeno Martín and Awah cited in Rodriguez Esteban, "El mapa del África Occidental Española," paragraph 43.
52. Rodriguez Esteban, paragraph 44.
53. Future Learn, "Living Ecosystems."
54. Berkson and Sulaiman, *Settled Wanderers*, 93–95. Used with permission from Sam Berkson, Mohammed Sulaiman, and Influx Press.

55. Hamza Lakhal and Zaim Allal, personal communication, June 29, 2022.
56. Irwin, "Derivative States," 9.
57. Irwin, 135.
58. Lakhal, "Bedouin Obsession," 25.
59. Ellison, "La amada Tiris," 89.
60. Hamza Lakhal, personal communication, September 18, 2018.
61. Boulay, "Habiter ou 'faire avec' le sable," 80.
62. Boulay, 81–85.
63. Boulay, 95.
64. Taine-Cheikh, "Following the Tracks of the Moors," 204.
65. Taine-Cheikh, 202.
66. Boulay and Gélard, "Vivre le sable," 14.
67. Personal communication.
68. Cymene Howe argues for wind to be considered as a key more-than-human actor in energy developments. See Howe, *Ecologics*.

Chapter 7

1. Moulud, interview by Joanna Allan, Algeria, October 11, 2019.
2. Moulud, interview by Joanna Allan, Algeria, October 7, 2019.
3. Held on October 12, 2019, Algeria.
4. Wenzel, *Disposition of Nature*.
5. Spice, "Fighting Invasive Infrastructures."
6. Spice.
7. Kinder, "Solar Infrastructure."
8. Kinder, 64.
9. Kinder, 67.
10. Kinder, 67.
11. Pseudonym.
12. Pseudonym.
13. Pseudonym.
14. Porges, "Environmental Challenges and Local Strategies," 8.
15. Saharawi Arab Democratic Republic, *First Indicative Nationally Determined Contribution*, 21.
16. Porges, "Environmental Challenges and Local Strategies," 8.
17. Brand et al., "Radical Climate Change Politics," 1.
18. Brand et al., 11–12.
19. Brooks, "Challenging Climate Colonialism."
20. Brooks.
21. Brooks.
22. Saharawi Arab Democratic Republic, *First Indicative Nationally Determined Contribution*, 32.
23. Mohammed, interview by Joanna Allan, Rabouni camp, Algeria, October 7, 2019.
24. Saharawi Arab Democratic Republic, *First Indicative Nationally Determined Contribution*, 16.
25. Mohammed, interview by Joanna Allan, Rabouni camp, Algeria, October 7, 2019.
26. Mohammed, interview.
27. For more on the role of gender equality in the SADR's external-facing nationalist discourses, see Allan, "Imagining Saharawi Women."
28. Calvert, "From 'Energy Geography' to 'Energy Geographies.'"
29. Hecht, *Radiance of France*; Labban, "Preempting Possibility"; Kale, *Electrifying India*.
30. Hecht, *Radiance of France*; Bouzarovski and Bassin, "Energy and Identity."
31. Power and Kirshner, "Powering the State."
32. Allan, "Imagining Saharawi Women."

33. Ruano Posada and Solana Moreno, "Strategy of Style."
34. Mohammed Saleh, interview by Joanna Allan, Rabouni camp, Algeria, October 7, 2019.
35. Dunlap, "More Wind Energy Colonialism(s) in Oaxaca?," 6.
36. Mohammed, interview by Joanna Allan, Rabouni camp, Algeria, October 7, 2019.
37. Howe, *Ecologics*, 41.
38. Saharawi Arab Democratic Republic, *First Indicative Nationally Determined Contribution*, 17.
39. Viswanathan, "Indigenous Engagement with Renewable Energy Projects."
40. Viswanathan.
41. Mohammed Saleh, interview by Joanna Allan, Rabouni camp, Algeria, October 7, 2019.
42. For more on the market economy in the camps, see Tavakoli, "Cultural Entrepreneurship of Sahrawi Refugees"; and Wilson, *Sovereignty in Exile*.
43. El Nour, *Towards a Just Agricultural Transition*.
44. Moustakbal, *Moroccan Energy Sector*.
45. Lennon, "Energy Transitions."
46. Lennon, 3.
47. Swanson and Buckley, "Chinese Solar Companies." Several researchers have highlighted the harsh working conditions and toxic environments associated with wind and solar farms more generally. See, for example, Dunlap, "Spreading 'Green' Infrastructural Harm"; Klinger, *Rare Earth Frontiers*; and Lennon, "Energy Transitions."
48. Lennon, "Energy Transitions," 3.
49. Lennon, 4.
50. Lennon, 8.
51. Dunlap, "Spreading 'Green' Infrastructural Harm," 20.
52. Mastini, "Degrowth."
53. Dunlap, "Spreading 'Green' Infrastructural Harm."
54. See Berkson and Sulaiman, *Settled Wanderers*, 60–65.
55. El Wali (pseudonym), interview by Joanna Allan, refugee camps, Algeria, October 8, 2019.
56. Porges, "Environmental Challenges and Local Strategies," 8.
57. Saharawi Arab Democratic Republic, *First Indicative Nationally Determined Contribution*, 34–35.
58. Saharawi Arab Democratic Republic, 15–16.
59. Saharawi Arab Democratic Republic, 15–16.
60. Saharawi Arab Democratic Republic, 15–16.
61. El Bechir Dedi, "Proyecto 'Water for All.'"
62. Ahmed, *Acceso al agua*.
63. El Wali, interview by Joanna Allan, refugee camps, Algeria, October 8, 2019.
64. Ibanga, "Renewable Energy Issues," 122–25.
65. Ibanga, 122–23.
66. Ibanga, 122.
67. Ibanga, 119.
68. Ibanga, 119.
69. Ibanga, 128.
70. Ibanga, 119; Lohmann, "White Climate, White Energy."
71. SADR Constitution quoted in Fadel, "Role of Natural Resources," 355.
72. My assertion here is based on a discussion I had with Saharawi youths at a natural resources workshop in Seville, Spain, in 2015.
73. Irwin, "Derivative States," 79.
74. Fadel, "Role of Natural Resources," 345.
75. See, for example, Fiddian-Qasmiyeh, *Ideal Refugee*; and Allan, *Silenced Resistance*.

76. Drury, "Rabouni Woven Panel."
77. Drury.
78. Fadel, "Role of Natural Resources," 345.
79. Irwin, "Derivative States," 81.
80. For more on these summer schools, see Irwin, "Derivative States," 140.
81. Stephan and Mundy, "A Battlefield Transformed," 31.
82. Taleb Brahim, interview by Joanna Allan, refugee camps, Algeria, October 11, 2019.
83. Taleb Brahim, interview.
84. Taleb Brahim, interview.
85. Taleb Brahim, interview.
86. Saharawi Arab Democratic Republic, *First Indicative Nationally Determined Contribution*, 32–33.
87. Porges, "Environmental Challenges and Local Strategies," 8.
88. Taleb Brahim, interview by Joanna Allan, refugee camps, Algeria, October 11, 2019.
89. Clarke and Brooks, *Archaeology of Western Sahara*, 197.
90. Taleb Brahim, interview by Joanna Allan, refugee camps, Algeria, October 11, 2019.
91. Saharawi Arab Democratic Republic, *First Indicative Nationally Determined Contribution*, 31.
92. Saharawi Arab Democratic Republic, 32.
93. Criado-Hernández, Moreno-Medina, and Dorta-Antequera, "Otra imagen del desierto," 8.
94. Criado-Hernández, Moreno-Medina, and Dorta-Antequera, 9.
95. Criado-Hernández, Moreno-Medina, and Dorta-Antequera, 10.
96. Coutard, "Placing Splintering Urbanism."
97. Nixon, *Slow Violence*, 112.

Chapter 8

1. Malainin Lakhal, personal conversation, March 17, 2021.
2. Yu et al., "Fertilizing Role of African Dust."
3. Scholz, *Political Solidarity*, 6.
4. Allan, "Imagining Saharawi Women"; San Martín, *Western Sahara*.
5. Kasanda, "Exploring Pan-Africanism's Theories," 182.
6. Fanon, *Les damnés de la terre*.
7. Scholz, *Political Solidarity*, 10.
8. Bens, "Retorno de un colonizador," 108.
9. Mayrata, *El imperio desierto*, 11.
10. Bens, *Relatos del Sáhara español*, 14.
11. Hughes, *Who Owns the Wind?*, 91. Hughes notes that Vázquez-Figueroa is also the author of Spain's only novel (to Hughes's knowledge at his time of writing in 2021) about wind turbines.
12. Vázquez-Figueroa, *Arena y viento*, 190.
13. Vázquez-Figueroa, 191.
14. Vázquez-Figueroa, 170.
15. Vázquez-Figueroa, 11.
16. Vázquez-Figueroa, 9.
17. Hsu-Ming, *Desert Passions*; Lucas, "Orientalism and Imperialism"; Saifuz Zaman, "Edification of Sir Walter Scott's Saladin," 42.
18. See especially Ellison, *Africa in the Contemporary Spanish Novel*, chapter 3.
19. Tajditi quoted in Moya, "La cuestión saharaui."
20. Moya, "La cuestión saharaui."
21. Ellison, *Africa in the Contemporary Spanish Novel*, 94.
22. Ellison, 77.

23. Mayrata, *El imperio desierto*, 53.
24. Mayrata, 357.
25. Mayrata, 70.
26. Mayrata, 152.
27. Mayrata, 90.
28. Reverte, *El médico de Ifni*, 24.
29. Ellison, *Africa in the Contemporary Spanish Novel*, 220.
30. Harris, *Weatherland*, 13.
31. Reverte, *El médico de Ifni*, 99–100.
32. Reverte, 212. Details on *Gatma* from an interview with Omar (pseudonym), occupied El Aaiún, September 8, 2020.
33. Reverte, 182.
34. Reverte, 76.
35. Reverte, 69.
36. Gerling, "El imperio de arena," 306.
37. Torbado, *El imperio de arena*, 110.
38. Torbado, 49.
39. López, "El viento como metáfora."
40. Mayrata, *El imperio desierto*, 53.
41. Romeo, *Discothèque*, 26.
42. Romeo, 27.
43. Romeo, 53–54.
44. Romeo, 54.
45. Romeo, 104.
46. Romeo, 119.
47. Romeo, 119–21.
48. Romeo, 120.
49. Romeo, 217.
50. Romeo, 23–24.
51. Bubisher "Quienes somos."
52. Bubisher.
53. Rodríguez, *Sand/Water*, 7.
54. Federico Guzmán, interview by Joanna Allan, Skype, July 23, 2018.
55. Available at https://www.youtube.com/watch?v=DBFCnURzdVw.
56. Federico Guzmán, interview by Joanna Allan, Skype, July 23, 2018.
57. Federico Guzmán, interview.
58. Federico Guzmán, interview.
59. Jabary Salamanca, "Palestinian 1968."

Conclusion

1. Pedro Sánchez (president of the Spanish government), letter to the King of Morocco, March 14, 2022.
2. Najla Mohammed, extract of a post from her personal Facebook page, March 17, 2022.
3. Mahmoud, interview by Joanna Allan, occupied El Aaiún, May 27, 2019.
4. Lohmann, "El cuestionamiento"; Daggett, *Birth of Energy*.
5. Stephan and Mundy, "A Battlefield Transformed," 31.
6. Dunlap and Laratte, "European Green Deal Necropolitics."
7. Lennon, "Energy Transitions."
8. See, for example, quotations of Sánchez in Ana Buil Demur, "Sánchez defiende en el congreso."

9. On green extractivism, see especially Dunlap and Brock, "Normalising Corporate Counterinsurgency"; Dunlap and Brock, "When the Wolf Guards the Sheep"; and Verweijen and Dunlap, "Evolving Techniques of Social Engineering."
10. Middle East Eye, "Morocco Bought Israel's Suicide Drones"; Middle East Monitor, "Morocco, Israel to Build 2 Drone Factories"; Atalayar, "Sabena y Lockheed."
11. Aroca, "Calima en la Moncloa."

Bibliography

Abram, David. *The Spell of the Sensuous: Perception and Language in a More-Than-Human World*. New York: Vintage Books, 2017 (first published 1996).

Africa Intelligence. "Lalla Noufissa Expands Her Dakhla Tourism Empire." October 31, 2019. https://www.africaintelligence.com/north-africa_business/2019/10/31/lalla-noufissa-expands-her-dakhla-tourism-empire,108379624-art.

———. "Western Sahara: Masen to Build Noor Solar Farm near El Argoub, Dakhla." August 24, 2020. https://www.africaintelligence.com/north-africa_business/2020/08/24/western-sahara-masen-to-build-noor-solar-farm-near-el-argoub-dakhla,109601298-bre.

Ahmed, Zahra R. *Acceso al agua en los campamentos de refugiados/as saharauis*. Zaragoza: Zaragoza Council, 2008.

Allan, Joanna. "Activist Ethics: The Need for a Nuanced Approach to Resistance Studies Field Research." *Resistance Studies* 3, no. 2 (2017): 89–121.

———. "Colonialism and Patriarchy in Afrohispanic Literature: The Resistance Writings of Lehdia Dafa and María Nsue Angüe." In *Confluencias culturales afro-hispanas: África, Latinoamérica y Europa*, edited by Dorothy Odartey-Wellington, 137–51. Leiden: Brill, 2018.

———. "Decolonizing Renewable Energy: Aeolian Aesthetics in the Poetry of Fatma Galia Mohammed Salem and Limam Boisha." *Bulletin of Hispanic Studies* 97, no. 4 (2020): 421–37.

———. "Imagining Saharawi Women: The Question of Gender in Polisario Discourse." *The Journal of North African Studies* 15, no. 2 (2010): 189–202.

———. "Nationalism, Resistance and Patriarchy: The Poetry of Saharawi Women." *Hispanic Research Journal* 12, no. 1 (2011): 78–89.

———. "The Saharawi 'Friendship Generation.'" In *The Literary Encyclopedia*, edited by Helen Rachel Cousins. Slough: Literary Dictionary Company Limited, 2017.

———. *Silenced Resistance: Women, Genderwashing, and Dictatorships in Western Sahara and Equatorial Guinea*. Madison: University of Wisconsin Press, 2019.

Allan, Joanna, and Hamza Lakhal. *Acting with Impunity: Morocco's Human Rights Violations in Western Sahara and the Silence of the International Community*. Oslo: Students and Academics International Assistance Fund (SAIH), 2015.

Allan, Joanna, Mahmoud Lemaadel, and Hamza Lakhal. "Oppressive Energopolitics in Africa's Last Colony: Energy, Subjectivities, and Resistance." *Antipode* 54, no. 1 (2022): 44–63.

Allan, Joanna, and Raquel Ojeda-García. "Natural Resource Exploitation in Western Sahara: New Research Directions." *Journal of North African Studies* 27, no. 6 (2021): 1107–36.

Al Montassir, Abdessamad. *Al Amakine*. Rabat: Independent Art Rome *Le Cube*, 2020.

———. Dossier Artistique. 2021.
———. *Galb'Echaouf*. Morocco, Western Sahara: Abdessamad El Montassir and Fondation IMéRA. Video, 0:18.
Alvarez de Benito, Gerardo. File of research documents, letters, and press cuttings on ways to stabilize sand dunes, held in the Spanish government's General Administrative Archives (AGA). March 21, 1972.
Amnesty International. "Morocco: Sahrawi Activist Raped by Moroccan Forces." November 2021. https://www.amnesty.org.uk/urgent-actions/sahrawi-activist-raped-moroccan-forces.
———. "Urgent Action: Sahrawi Activist Raped by Moroccan Forces." November 30, 2021. https://www.amnesty.org.uk/files/2021-12/FI03321_2.pdf?VersionId=YHMb7zCKEk1skOuY.zD1IPwymOfBhQna.
Amoko, Apollo. "The 'Missionary Position' and the Postcolonial Polity, or, Sexual Difference in the Field of Kenyan Colonial Knowledge." *Callaloo* 24, no. 1 (2001): 310–24.
Anderson, Katherine. "The Weather Ship: Networks, Disasters, and Imaginaries after 1945." In *Weather, Climate, and the Geographical Imagination*, edited by Martin Mahony and Samuel Randalls, 67–92. Pittsburgh: University of Pittsburgh Press, 2020.
Angel, James. "New Municipalism and the State: Remunicipalising Energy in Barcelona, from Prosaics to Process." *Antipode* 49, no. 3 (2021): 557–76.
Apter, David E. *Rethinking Development: Modernization, Dependency, and Postmodern Politics*. Thousand Oaks, CA: Sage Publications, 1987.
Aravind, Sakshi. "The Many Entanglements of Capitalism, Colonialism and Indigenous Environmental Justice: How Can Legal and Other Challenges Bring a Halt to the Destruction of Indigenous Lands by Global Corporations?" *Soundings* 2021, no. 78 (2021): 64–80.
Aroca, Javier. "Calima en la Moncloa." *El Diario*. Accessed June 23, 2022. https://www.eldiario.es/andalucia/desdeelsur/calima-moncloa-sahara-pedro-sanchez_132_8844801.html.
Asouzu, Innocent I. "Ibuanyidanda (Complementary Reflection), Communalism and Theory Formulation in African Philosophy." *Thought and Practice: A Journal of the Philosophical Association of Kenya* 3, no. 2 (2011): 9–34.
Association for the Monitoring of the Resources and the Protection of the Environment (AMRPENWS). "Environmental Issues in Western Sahara." August 29, 2016. sahararesources.org/environmental-issues-in-western-sahara. Site defunct.
Atalayar. "Sabena y Lockheed se ponen al servicio de la defensa aérea marroquí." Accessed December 12, 2023. https://www.atalayar.com/en/articulo/economy-and-business/sabena-and-lockheed-serve-moroccan-air-defence-industry/20220419133320156048.html.
Athreya, Venkatesh. "Perestroika and the Third World: The Changing Status of the Concept of 'Neocolonialism.'" *Social Sciences* 7/8 (1989): 28–36.
Atlas S.A. combustibles y lubrificantes. *Dossier: Factorías de servicio en El Aaiún y Aaiún Playa*. November 1961.
Awah, Bahia Mahmud. "Literatura oral y transmisión en el Sáhara." *Quaderns de la Mediterrània* 13 (2010): 207–10.
———. *Tiris: Rutas literarias*. Madrid: Última Línea, 2016.
Azurara, Gomes Eannes de. *The Chronicle of the Discovery and Conquest of Guinea, Volume 1*. Charleston, SC: Bibliolife, 1896.
Baba Miské, Ahmed. *Front Polisario, L'âme d'un peuple*. Paris: Editions Rupture, 1978.
Baepler, Paul. "The Barbary Captivity Narrative in American Culture." *Early American Literature* 39, no. 2 (2004): 217–46.
Bandt, Ros. "Hearing Australian Identity: Sites as Acoustic Spaces, an Audible Polyphony." *Resonate Magazine*. August 27, 2020. https://www.australianmusiccentre.com.au/article/hearing-australian-identity.

Bankoff, Greg. "Winds of Colonisation: The Meteorological Contours of Spain's Imperium in the Pacific 1521–1898." *Environment and History* 12, no. 1 (2006): 65–88.
Baptista, Idalina. "Maputo: Fluid Flows of Power and Electricity—Prepayment as Mediator of State-Society Relationships." In *Energy, Power, and Protest on the Urban Grid: Geographies of the Electric City*, edited by Andrés Luque-Ayala and Jonathan Silver, 112–32. London: Routledge, 2016.
Bárbulo, Tomás. "Batalla por el fuerte de Villa Cisneros." El País. July 20, 2004. https://elpais.com/diario/2004/07/21/ultima/1090360801_850215.html.
———. *La historia prohibida del Sáhara español*. Barcelona: Ediciones Destino, 2002.
Barker, Adam J. "Locating Settler Colonialism." *Journal of Colonialism and Colonial History* 13, no. 3 (2012). https://muse.jhu.edu/article/491173.
Barona Castañeda, Claudia. "Memorias de una resistencia. La otra historia del Sahara occidental." *Les Cahiers d'ÉMAM* 24–25 (2015). https://journals.openedition.org/emam/859.
Bassett, Thomas J. "Mobile Pastoralism on the Brink of Land Privatization in Northern Côte d'Ivoire." *Geoforum* 40 (2009): 756–66.
BBC World Service. "The Sounds of the Namib Desert." 2018. https://player.fm/series/the-compass-1301444/the-sounds-of-the-namib-desert.
Beck, John. "Without Form and Void: The American Desert as Trope and Terrain." *Neplantla: Views from South* 2, no. 1 (2001): 63–83.
Bedjaoui, Mohammed. *Terra nullius, "droits" historiques et autodétermination*. The Hague: Sijthoff, 1975.
Benítez, Cristóbal. "La arena huye del agua: Camino de Tombuctú." In *Relatos del Sáhara español*, edited by Ramón Mayrata, 23–30. Madrid: Clan, 2005.
Bennis, Amine. "Morocco's Contemporary Diplomacy as Middle Power." *Journal of International Affairs*. August 10, 2019. https://jia.sipa.columbia.edu/online-articles/moroccos-contemporary-diplomacy-middle-power.
Bens, Francisco. "Retorno de un colonizador." In *Relatos del Sáhara español*, edited by Ramón Mayrata, 107–10. Madrid: Clan, 2005.
———. "El sabor de la planta de los pies." In *Relatos del Sáhara español*, edited by Ramón Mayrata, 99–106. Madrid: Clan, 2005.
Berkson, Sam. "Voices of a Lost Homeland: The Poetry of Western Sahara." Middle East Eye. September 23, 2020. https://www.middleeasteye.net/discover/western-sahara-algeria-badi-poetry-exile?fbclid=IwAR1urbWAvYFlKonM8vSu83waq3oJcQdhQGVsBYiAKEIs5r8zdLrGKjjVST4.
Berkson, Sam, and Mohamed Sulaiman. *Settled Wanderers: The Poetry of Western Sahara*. London: Influx Press, 2015.
Black, Brian C. *Crude Reality: Petroleum in World History*. Lanham: Rowman & Littlefield, 2012.
Black, Hector. *A Narrative of the Shipwreck of the British Brig "Surprise" of Glasgow, John William Ross, Master, on the Coast of Barbary on 28 December 1815, and Subsequent Captivity of the Passengers and Crew by the Arabs until Ransomed by the Worshipful Company of Ironmongers*. London: Gye and Balne, 1817.
Blackmore, Josiah. *Manifest Perdition: Shipwreck Narrative and the Perdition of Empire*. Minneapolis: University of Minnesota Press, 2002.
———. *Moorings: Portuguese Expansion and the Writing of Africa*. Minneapolis: University of Minnesota Press, 2008.
Boast, Hannah Kate. *Hydrofictions: Water, Power and Politics in Israeli and Palestinian Literature*. Edinburgh: Edinburgh University Press, 2020.
Boisha, Limam. *Ritos de jaima*. Madrid: Editorial Bubisher, 2012.
Boletsi, Maria. *Barbarism and Its Discontents*. Stanford: Stanford University Press, 2013.
Bonelli, Emilio. *El Sahara: Descripción geográfica, comercial y agrícola desde Cabo Bojador a Cabo Blanco, viajes al interior, habitantes del desierto y consideraciones generales*. Madrid:

Ministerio de Fomento, 1887. https://www.usc.es/export9/sites/webinstitucional/gl/institutos/ceso/descargas/Biblio_Bonelli_Sahara-descripcin.geogrfica_1887.pdf.

Boulakhbar, Mouaad, Badr Lebrouhi, Tarik Kousksou, Slimane Smouh, Abdelmajid Jamil, Mohamed Maaroufi, and Malika Zazi. "Towards a Large-Scale Integration of Renewable Energies in Morocco." *Journal of Energy Storage* 32 (2020): 1–17.

Boulay, Sébastien. "Habiter ou 'faire avec' le sable: Domestiquer l'espace et la matière en contexte saharien." *Techniques & Culture* 61 (2013): 76–99.

———. "'Returnees' and Political Poetry in Western Sahara: Defamation, Deterrence and Mobilisation on the Web and Mobile Phones." *Journal of North African Studies* 21, no. 4 (2016): 667–86.

Boulay, Sébastien, and Marie-Luce Gélard. "Vivre le sable: Une introduction." *Techniques & Culture* 61 (2013): 10–27.

Bouzarovski, Stefan, and Mark Bassin. "Energy and Identity: Imagining Russia as a Hydrocarbon Superpower." *Annals of the Association of American Geographers* 101, no. 4 (2011): 783–94.

Boyer, Dominic. *Energopolitics: Wind and Power in the Anthropocene*. Durham: Duke University Press, 2019.

———. "Energopower: An Introduction." In *Energy Humanities: An Anthology*, edited by Imre Szeman and Dominic Boyer, 184–204. Baltimore: John Hopkins University Press, 2017.

Branche, Jerome. "Inscribing Contact: Zurara, the Africans, and the Discourse of Colonialism." *PALARA: Publication of the Afro-Latin/American Research Association* 5 (2001): 6–19.

Brand, Ulrich, Nicola Bullard, Edgardo Lander, and Tadzio Mueller. "Radical Climate Change Politics in Copenhagen and Beyond: From Criticism to Action?" In *Critical Currents*, edited by Ulrich Brand, Nicola Bullard, Edgardo Lander, and Tadzio Mueller, 9–16. Uppsala: Dag Hammarskjöld Foundation, 2009.

Brew-Hammond, Abeeku. "Energy Resources in Africa: Challenges Ahead." *Energy Policy* 38, no. 5 (2010): 2291–301.

Brooks, Nick. "Challenging Climate Colonialism in North Africa—a New Indicative NDC for Western Sahara." Nick Brooks. November 15, 2021. https://nickbrooks.wordpress.com/2021/11/15/challenging-climate-colonialism-in-north-africa-a-new-indicative-ndc-for-western-sahara/.

Brown, Roger, and Albert Gilman. "The Pronouns of Power and Solidarity." In *Style in Language*, edited by Thomas A. Sebeok, 253–76. Cambridge, MA: MIT Press, 1960.

Bruchac, Margaret M. "Indigenous Knowledge and Traditional Knowledge." In *Encyclopedia of Global Archaeology*, edited by Claire Smith, 3814–24. New York: Springer Science and Business Media, 2014.

Bubisher. "Quienes somos." Accessed June 22, 2022. http://www.bubisher.org/quienessomos/.

Buil Demur, Ana. "Sánchez defiende en el congreso la nueva postura sobre el Sáhara y las medidas para bajar la luz." Euronews. Accessed June 23, 2022. https://es.euronews.com/2022/03/30/sanchez-defiende-en-el-congreso-la-nueva-postura-sobre-el-sahara-y-las-medidas-para-bajar-.

Caffarena Aceña, Vicente (Road Engineer). *Informe sobre los daños causados en la ciudad de Sidi Ifni por la riada de la noche del 11 al 12 de noviembre de 1967*. Malaga, November 26, 1967.

Calvert, Kirby. "From 'Energy Geography' to 'Energy Geographies': Perspectives on a Fertile Academic Borderland." *Progress in Human Geography* 40, no. 1 (2016): 105–25.

Campos Serrano, Alicia, and Violeta Trasosmontes. "Recursos naturales y segunda ocupación colonial del Sahara española." *Les Cahiers d'ÉMAM* 24–25 (2015). https://journals.openedition.org/emam/819?lang=en.

Camprubí, Lino. "Resource Geopolitics: Cold War Technologies, Global Fertilizers, and the Fate of Western Sahara." *Technology and Culture* 56, no. 3 (2015): 676–703.

Cartwright, Mark. "Prince Henry the Navigator." *World History Encyclopedia*. August 4, 2021. https://www.worldhistory.org/Prince_Henry_the_Navigator/.

Cervera, Julio. "Expedición al Sáhara." *Revista de geografía comercial española* 25–30, Year 2 (1886): 1–110.

Chapman, Chelsea, Arthur Mason, Devon Reeser, Rupp Stephanie, and Sarah Strauss. "Conversation 2: Technology, Meaning, Cosmology." In *Cultures of Energy: Power, Practices, Technologies*, edited by Sarah Strauss, Stephanie Rupp, and Thomas Love, 139–44. Walnut Creek, CA: Left Coast Press, 2013.

Chikowero, Moses. "Subalternating Currents: Electrification and Power Politics in Bulawayo, Colonial Zimbabwe, 1894–1939." *Journal of Southern African Studies* 33, no. 2 (2007): 287–306.

Clarke, Joanna, and Nick Brooks. *The Archaeology of Western Sahara: A Synthesis of Fieldwork, 2002 to 2009*. Oxford: Oxbow Books, 2018.

Cochelet, Charles. *Narrative of the Shipwreck of the Sophia, on the 30th of May 1819*. London: Sir R. Phillips and Co., 1822.

Cock, Simon. *The Narrative of Robert Adams: An American Sailor Who Was Wrecked on the Western Coast of Africa, in the Year 1810, Was Detained Three Years in Slavery by the Arabs of the Great Desert, and Resided Several Months in the City of Tombuctoo*. Boston: Wells and Lilly, 1816.

Communications engineers on behalf of the General Directorate for African Provinces and Posts. *Proyecto de comunicaciones radioeléctricas del Sáhara*. Madrid, January 1961.

Congost Tor, Josep. "Estudio ornitológico de la región de Seguiat-El-Hamra, Sahara español, en abril de 1973." *Miscelánea Zoológica* 111, no. 5 (1976): 195–208.

Corell, Hans. *Letter Dated 29 January 2002 from the Under-Secretary-General for Legal Affairs, the Legal Counsel, Addressed to the President of the Security Council*. United Nations Security Council, 2002. http://www.arso.org/UNlegaladv.htm.

Correale, Francesco, and Alberto López Bargados. "Rashōmon au Sahara occidental: Perspectives, Contradictions et défis dans l'interprétation du conflit de 1956–1958." In *Culture et politique dans l'Ouest saharien: Arts, activisme et état dans un espace de conflits: (Algérie, Mali, Maroc, Mauritanie, Sahara occidental)*, edited by Sébastien Boulay and Francesco Freire, 211–41. Paris: Igé, 2017.

Coutard, Olivier. "Placing Splintering Urbanism: Introduction." *Geoforum* 39, no. 6 (2008): 1815–20.

Criado-Hernández, Constantino, Claudio Moreno-Medina, and Pedro Dorta-Antequera. "Otra imagen del desierto. El clima del antiguo Sahara español entre 1882 y 1890." *Ería* 1 (2017): 5–19.

Crook, Martin, and Damien Short. "Developmentalism and the Genocide–Ecocide Nexus." *Journal of Genocide Research* 23, no. 2 (2021): 162–88.

Crooks, Ed, and Patrick McGee. "GE and Siemens: Power Pioneer Flying Too Far from the Sun." *Financial Times*, November 12, 2017. https://www.ft.com/content/fc1467b8-c601-11e7-b2bb-322b2cb39656.

Cudd, Ann E. "How to Explain Oppression: Criteria of Adequacy for Normative Explanatory Theories." *Philosophy of Social Sciences* 35, no. 1 (2005): 20–49.

Dachverband der Kritischen Aktionärinnen and Aktionäre. *Counterproposals Concerning the Annual Shareholders' Meeting of Siemens AG on February 1, 2017*. Siemens. February 1, 2017. https://www.siemens.com/investor/pool/en/investor_relations/events/annual_shareholders_meeting/2017/gegenantraege_e_hv2017_170113.pdf.

Daggett, Cara. *The Birth of Energy: Fossil Fuels, Thermodynamics, and the Politics of Work*. Durham: Duke University Press, 2019.

Dall'omo, Sabine. "Opinion: We Are All Empowered by Access to Electricity." Independent Online. March 19, 2018. https://www.iol.co.za/business-report/opinion/opinion-we-are-all-empowered-by-access-to-electricity-13907775.

D'Almonte, Enrique. "El Sahara español: Esbozo orográfico e indicaciones geológicas." *Journal of Colonisation: Industry, Commerce, Moral and Material Interests* 10 (January 10, 1917): 59–74.

Davila, Carl. "East Winds and Full Moons: Ramal Al-Māya and the Peregrinations of Love-Poetry Images." *Journal of North African Studies* 19, no. 1 (2014): 7–26.

DeBlieu, Jan. *Wind: How the Flow of Air Has Shaped Life, Myth, and the Land*. Boston: Mariner Books, 1999.

DeBoom, Meredith. "Climate Necropolitics: Ecological Civilization and the Distributive Geographies of Extractive Violence in the Anthropocene." *Annals of the American Association of Geographers* 111, no. 3 (2021): 900–912.

Díaz de Villegas, Jose. *Proyecto de construcciones e instalaciones de producción y distribución de energía eléctrica en l Aaiún*. November 28, 1960. Held in the Spanish government's General Administrative Archives (AGA).

Diego Aguirre, José Ramon. *Historia del Sahara español*. Madrid: Kayededa, 1988.

Domenech Lafuente, Ángel. *Algo sobre Río de Oro*. Madrid: Gobierno Político Militar del África Occidental Española, 1946.

Drury, Mark. "Anticolonial Irredentism: The Moroccan Liberation Army and Decolonisation in the Sahara." *Journal of North African Studies* (2022): 1–27. https://www.tandfonline.com/doi/full/10.1080/13629387.2022.2056449#.

———. "Disorderly Histories: An Anthropology of Decolonization in Western Sahara." PhD diss., City University of New York (CUNY), 2018.

———. "Global Future and Government Towns: Phosphates and the Production of Western Sahara as a Space of Contention." *Arab World Geographer* 16, no. 1 (2013): 101–24.

———. "On the Rabouni Woven Panel." MoMA. Insecurities: Tracing Displacement and Shelter. February 1, 2017. https://medium.com/insecurities/on-the-rabouni-woven-panel-7004fc8755c8.

Dunlap, Alexander. "Does Renewable Energy Exist? Fossil Fuel+ Technologies and the Search for Renewable Energy." In *A Critical Approach to the Social Acceptance of Renewable Energy Infrastructure*, edited by Susana Batel and David Rudolph, 83–102. London: Palgrave, 2021.

———. "More Wind Energy Colonialism(s) in Oaxaca? Reasonable Findings, Unacceptable Development." *Energy Research & Social Science* 82 (2021): 1–8.

———. *Renewing Destruction: Wind Energy Development, Conflict and Resistance in a Latin American Context*. Lanham, MD: Rowman & Littlefield, 2019.

———. "Spreading 'Green' Infrastructural Harm: Mapping Conflicts and Socio-ecological Disruptions within the European Union's Transnational Energy Grid." *Globalizations* 20, no. 6 (2021): 1–25.

Dunlap, Alexander, and Martín Correa Arce. "'Murderous Energy' in Oaxaca, Mexico: Wind Factories, Territorial Struggle and Social Warfare." *Journal of Peasant Studies* 49, no. 2 (2021): 455–80.

Dunlap, Alexander, and Andrea Brock. "Normalising Corporate Counterinsurgency: Engineering Consent, Managing Resistance and Greening Destruction around the Hambach Coal Mine and Beyond." *Political Geography* 62 (2017): 33–47.

———. "When the Wolf Guards the Sheep: The Industrial Machine through Green Extractivism in Germany and Mexico." In *Energies beyond the State: Anarchist Political Ecology and the Liberation of Nature*, edited by Jennifer Mateer, Simon Springer, Martin Locret-Collet, and Maleea Acker, 91–123. Lanham, MD: Rowman and Littlefield, 2021.

Dunlap, Alexander, and Louis Laratte. "European Green Deal Necropolitics: Exploring 'Green' Energy Transition, Degrowth and Infrastructural Colonization." *Political Geography* 97 (2022): 1–17.

Eau du Maroc. "Réseau electrique au Sahara marocain." 2014. Accessed June 21, 2019. http://www.eaudumaroc.com/2017/04/reseau-electrique-au-sahara-marocain_88.html.

El Bechir Dedi, Moulay. "Proyecto 'Water for All' por el Sahara occidental utilizando las energía renovables." July 11, 2020. Lausanne, Switzerland.

Ellison, Mahan L. *Africa in the Contemporary Spanish Novel, 1990–2010*. London: Lexington Books, 2021.

———. "'La amada Tiris, tierra de nuestros abuelos': The Affective Space of the Sahara in Hispano-Sahrawi Literature." *CELAAN: Review of the Center for the Studies of the Literatures and Arts of North Africa* 15, no. 2–3 (2018): 73–102.

El Nour, Saker. *Towards a Just Agricultural Transition in North Africa*. Amsterdam: Transnational Institute (TNI), 2021. https://longreads.tni.org/towards-a-just-agricultural-transition-in-north-africa#note59.

Endfield, Georgina. "The British Women's Emigration Association and Climate(S) of South Africa." In *Weather, Climate, and the Geographical Imagination: Placing Atmospheric Knowledges*, edited by Martin Mahony and Samuel Randalls, 132–51. Pittsburgh: University of Pittsburgh Press, 2020.

Engineer of Pathways, Canals and Ports, General Government of the Sahara Province, Provincial Service for Public Works. *Report: Extension of the Smara Power Station and Its Distribution Lines*. El Aaiún: General Government of the Sahara Province, November 1972.

Enns, Charid, and Brock Bersaglio. "On the Coloniality of 'New' Mega-infrastructure Projects in East Africa." *Antipode* 52, no. 1 (2020): 101–23.

Environmental Justice Atlas. "Wind Power Plants in Occupied Territories of Western Sahara." February 5, 2022. https://ejatlas.org/print/wind-power-plants-in-occupied-territories-of-western-sahara.

Fadel, Kamal. "The Role of Natural Resources in the Building of an Independent Western Sahara." *Global Change, Peace and Security* 27, no. 3 (2015): 345–59.

Fairhead, James, Melissa Leach, and Ian Scoones. *Greengrabbing: A New Appropriation of Nature?* London: Routledge, 2013.

Fanon, Frantz. *Les damnés de la terre*. Paris: Maspero, 1979.

Fenton, Lori. "Sand Waves in the Desert. Or Pet Peeves and Deciphering Climate Change in the Solar System," Planetary Society. February 21, 2014. https://www.planetary.org/articles/0219-sand-waves-in-the-desert.

Fernández-Armesto, Felipe. *Before Columbus: Exploration and Colonization from the Mediterranean to the Atlantic, 1229–1492*. Philadelphia: University of Pennsylvania Press, 1987.

Fernández Camporro, Alicia. "The King's Speech. An Analysis of the Institutional Discourse around the Development of Renewable Energy Infrastructure in Western Sahara." *Journal of North African Studies* (forthcoming).

Fiddian-Qasmiyeh, Elena. *The Ideal Refugees: Gender, Islam, and the Sahrawi Politics of Survival*. New York: Syracuse University Press, 2014.

Flikke, Rune. "Matters That Matter: Air and Atmosphere as Material Politics in South Africa." In *Anthropos and the Material*, edited by Penny Harvey, Christian Krohn-Hansen, and Knut G. Nustad, 179–96. Durham: Duke University Press, 2019.

Flores Morales, Ángel. "Tipos y costumbres del Sáhara español." *África* 95 (1949): 407–10.

Foucault, Michel. *The History of Sexuality, Volume 1: An Introduction*. New York: Pantheon, 1978.

Foxhall, Katherine. "Interpreting the Tropical Atlantic Climate: Diaries from the Mid-Nineteenth-Century Australian Voyage." *Weather, Climate, and Society* 2 (2010): 91–102.

Franquesa, Jaume. *Power Struggles: Dignity, Value, and the Renewable Energy Frontier in Spain*. Bloomington: Indiana University Press, 2018.

Freedom House, "Freedom in the World 2022: Western Sahara." 2022. Accessed November 15, 2022. https://freedomhouse.org/country/western-sahara/freedom-world/2022.

Frontline Defenders. "House Arrest and Physical Assault of Woman Human Rights Defender Sultana Khaya." February 15, 2021. https://www.frontlinedefenders.org/en/case/house-arrest-and-physical-assault-woman-human-rights-defender-sultana-khaya.

———. "Woman Human Rights Defender Sultana Khaya Was Sexually Assaulted." May 14, 2021. https://www.frontlinedefenders.org/en/case/woman-human-rights-defender-sultana-khaya-was-sexually-assaulted.

Future Learn. "Living Ecosystems." Accessed June 30, 2022. https://www.futurelearn.com/info/courses/remaking-nature/0/steps/16724.

Galvan Balaguer, Jose. *Memoria: Proyecto de alumbrado eléctrico en el fuerte de la Factoría de Villas Cisneros y en la península de Rio de Oro*. Santa Cruz de Tenerife, 1913. Held in the Spanish government's General Administrative Archives (AGA).

García-Baquero, Manuel. "Cartografía militar Africana-Española." *Archivos del Instituto de Estudios Africanos*, no. 80 (1966): 21–49.

Geisler, Charles. "New Terra Nullius Narratives and the Gentrification of Africa's 'Empty Lands.'" *Journal of World-Systems Research* 18, no. 1 (2012): 15–29.

General Court of the European Union. "The General Court Annuls the Council Decisions Concerning, First, the Agreement between the European Union and Morocco Amending the Tariff Preferences Granted by the European Union to Products of Moroccan Origin and, Second, the Sustainable Fisheries Partnership Agreement." News release. September 29, 2021.

Gerling, David Ross. "El imperio de arena." *World Literature Today* 73, no. 2 (1999): 306.

Ghayour, Sabrina. *Sirocco: Fabulous Flavours from the East*. London: Mitchell Beazley, 2016.

Gilbert, Jérémie. *Nomadic Peoples and Human Rights*. London: Routledge, 2014.

Goldthau, Andreas. "Energy Diplomacy in Trade and Investment of Oil and Gas." In *Global Energy Governance: The New Rules of the Game*, edited by Andreas Goldthau and Jan Martin Witte, 25–48. Washington, DC: Brookings Press, 2010.

Government Delegate. *Permission Document*. Villa Cisneros: January 23, 1963.

Greene, Tommy, and Eoghan Gilmartin. "How Europe's Migrant Crisis Reached the Canary Islands." Novara Media. November 26, 2021. https://novaramedia.com/2021/11/16/how-europes-migrant-crisis-reached-the-canary-islands/.

Greenpeace. *Exporting Exploitation: How Retired Eu Fishing Vessels Are Devastating West African Fish Stocks and Undermining the Rights of Local People*. Stockholm: Greenpeace, 2013.

"Greenwashing." Accessed May 15, 2020. http://www.greenwashingindex.com/about-greenwashing/.

Grundy, Alice. "Octopus Invests in Morocco-UK Power Project Xlinks as It Signs Strategic Partnership." Current. May 12, 2022. https://www.current-news.co.uk/news/octopus-invests-in-morocco-uk-power-project-xlinks-as-it-signs-strategic-partnership.

Guinea López, Emilio. *España y el desierto: Impresiones saharianas de un botánico español*. Madrid: Instituto de Estudios Políticos, 1945.

Hagen, Erik, and Mario Pfeifer. *Profit over Peace in Western Sahara: How Commerical Interests Undermine Self-Determination in the Last Colony in Africa*. Berlin: Sternberg Press, 2018.

Haidar, Larosi. "La juventud de 'Irifi': Visiones y sentimientos de jóvenes saharauis." In *Nosotras, ustedes y ellos: Espacios, interacciones y exclusiones durante el periodo colonial y poscolonial en el Norte de África*, edited by Valeria Aguiar Bobet, 265–93. Las Palmas de Gran Canaria: Idea Ediciones, 2022.

Haila, Yrjö. "Beyond the Nature-Culture Dualism." *Biology and Philosophy* 15 (2000): 155–75.

Hamouchene, Hamza. "Desertec: The Renewable Energy Grab?" New Internationalist. March 1, 2015. https://newint.org/features/2015/03/01/desertec-long.

Harris, Alexandra. *Weatherland: Writers and Artists under English Skies*. London: Thames and Hudson, 2015.

Harris, Cole. "How Did Colonialism Dispossess? Comments from an Edge of Empire." *Annals of the American Association of Geographers* 94, no. 1 (2004): 165–82.

Harrison, Conor. "The American South: Electricity and Race in Rocky Mount, North Carolina, 1900–1935." In *Energy, Power and Protest on the Urban Grid*, edited by Andrés Luque-Ayala and Jonathan Silver, 21–44. London: Routledge, 2016.

Hassan, Khalid. "Morocco Flirts with Ethiopia Amid Stalled Nile Dam Talks." Al Monitor. October 2, 2021. https://www.al-monitor.com/originals/2021/10/morocco-flirts-ethiopia-amid-stalled-nile-dam-talks.

Hecht, Gabrielle. *The Radiance of France: Nuclear Power and National Identity after World War II*. Cambridge, MA: MIT Press, 1998.

Hernández Pacheco, Eduardo and Francisco. "El viento en el desierto." In *Relatos del Sáhara español*, edited by Ramón Mayrata, 205–10. Madrid: Clan Editorial, 2005.

———. *Sahara español*. Madrid: University of Madrid, 1942.

Hidalgo de Cisneros, Ignacio. "Encuentro con Saint-Exupéry." In *Relatos del Sáhara español*, edited by Ramón Mayrata, 117–24. Madrid: Clan Editorial, 2005.

Hoene, Christin. "The Sounding City: Soundscapes and Urban Modernity in Amit Chaudhuri's Fiction." In *Re-inventing the Postcolonial (in the) Metropolis*, edited by Annika Bauer and Cecile Sandten, 363–78. Leiden: Brill, 2016.

Hopman, Marieke. "Covert Qualitative Research as a Method to Study Human Rights under Authoritarian Regimes." *Journal of Human Rights Practice* 13, no. 3 (2022): 1–17.

Howarth, David. *Discourse: Concepts in the Social Sciences*. Buckingham: Open University Press, 2000.

Howe, Cymene. *Ecologics: Wind and Power in the Anthropocene*. Durham: Duke University Press, 2019.

Howe, Cymene, and Dominic Boyer. "Aeolian Politics." *Distinktion: Scandinavian Journal of Social Theory* 16, no. 1 (2015): 31–48.

Hsu-Ming, Teo. *Desert Passions: Orientalism and Romance Novels*. Austin: University of Texas Press, 2012.

Hughes, David McDermott. *Energy without Conscience: Oil, Climate Change and Complicity*. Durham: Duke University Press, 2017.

———. *Who Owns the Wind? Climate Crisis and the Hope of Renewable Energy*. London: Verso, 2021.

Huh, Sookyeon. "Title to Territory in the Post-Colonial Era: Original Title and Terra Nullius in the ICJ Judgements in Cases Concerning Litigan/Sipadan (2002) and Pedra Branca (2008)." *European Journal of International Law* 26, no. 3 (2015): 709–25.

Human Rights Clinic of Cornell Law School and Clinic on Fundamental Rights of the Université de Caen Basse-Normandie. *Report on the Kingdom of Morocco's Violations of the International Covenant on Economic Social and Cultural Rights in the Western Sahara*. New York: UN Committee on Economic, Social and Cultural Rights, 2015. https://www.ecoi.net/en/file/local/1188166/1930_1461234104_int-cescr-css-mar-21582-e.pdf.

Human Rights Watch. *Keeping It Secret: The United Nations Operation in the Western Sahara*. New York: Human Rights Watch, 1995. https://www.hrw.org/legacy/reports/1995/Wsahara.htm.

———. "Morocco/Western Sahara: No Action on Police Beating of Rights Worker Dead-End Investigations Fuel Impunity." May 15, 2012. https://www.hrw.org/news/2012/05/15/morocco/western-sahara-no-action-police-beating-rights-worker.

Hunold, Christian, and Steven Leitner. "'Hasta La Vista, Baby!' The Solar Grand Plan, Environmentalism, and Social Constructions of the Mojave Desert." *Environmental Politics* 20, no. 5 (2011): 687–704.

Ibanga, Diana-Abasi. "Renewable Energy Issues in Africa Contexts." *Relations: Beyond Anthropocentrism* 6, no. 1 (2018): 117–33.

Ingold, Tim. "Earth, Sky, Wind, and Weather." *Journal of the Royal Anthropological Institute* 13 (2007): 19–38.

———. "The Eye of the Storm: Visual Perception and the Weather." *Visual Studies* 20, no. 2 (2005): 97–104.

———. "Footprints through the Weather-World: Walking, Breathing, Knowing." *Journal of the Royal Anthropological Institute* 16 (2010): 121–39.

International Court of Justice. "Western Sahara Advisory Opinion of 16 October 1975." October 16, 1975. https://www.icj-cij.org/public/files/case-related/61/061-19751016-ADV-01-00-EN.pdf.

Irish Times. "Fortress Europe or Fair Trade." October 18, 2005. https://www.irishtimes.com/opinion/fortress-europe-or-fair-trade-1.507015.

Irving, James. "Copy of Mr. James Irving's Journal When Shipwrecked on the Coast of Barbary, 1789." In *Slave Captain: The Career of James Irving in the Liverpool Slave Trade*, edited by Suzanne Schwarz, 82–108. Wrexham, Clwyd: Bridge Books, 1995.

Irwin, Randi. "Derivative States: Property Rights and Claims-Making in a Contested Territory." PhD diss., New School for Social Research, 2019.

Isidoros, Konstantina. *Nomads and Nation-Building in the Western Sahara: Gender, Politics and the Sahrawi*. London: I.B. Tauris, 2018.

———. "The Silencing of Unifying Tribes: The Colonial Construction of Tribe and Its 'Extraordinary Leap' to Nascent Nation-State Formation in Western Sahara." *Journal of the Anthropological Society of Oxford* 7, no. 2 (2015): 168–90.

Jabary Salamanca, Omar. "The Palestinian 1968: Struggles for Dignity and Solidarity." Rektoverso. May 31, 2018. https://www.rektoverso.be/artikel/struggles-for-dignity-and-solidarity-.

———. "Unplug and Play: Manufacturing Collapse in Gaza." *Human Geography* 4, no. 1 (2011): 22–37.

———. *When Settler Colonialism Becomes 'Development': 'Fabric of Life' Roads and the Spatialities of Development in the Palestinian West Bank*. Birzeit: Birzeit University, Centre for Development Studies, 2012.

Jackson, James Grey. *An Account of the Empire of Morocco and the Districts of Suse and Tafilelt*. London: W. Bulmer and Co., 1810.

Kale, Sunila S. *Electrifying India: Regional Political Economies of Development*. Stanford, CA: Stanford University Press, 2014.

Kasanda, Albert. "Exploring Pan-Africanism's Theories: From Race-Based Solidarity to Political Unity and Beyond." *Journal of African Cultural Studies* 28, no. 2 (2016): 179–95.

Kaur, Raminder. "Southern Spectrums: The Raw to the Smooth Edges of Energopower." In *Ethnographies of Power: A Political Anthropology of Energy*, edited by Tristan Loloum, Simone Abram, and Nathalie Ortar, 24–51. New York: Berghahn, 2021.

Kenneally, Tim. "Russell Crowe Movie 'In Sand and Blood' Sparks Copyright Lawsuit." The Wrap. December 17, 2015. https://www.thewrap.com/russell-crowe-movie-in-sand-and-blood-sparks-copyright-lawsuit/.

Khatib, Neimat. *Morocco Country Profile*. Rome: Res4Med, 2018. https://www.res4med.org/wp-content/uploads/2018/06/Country-profile-Marocco-2.pdf.

Kinder, Jordan B. "Solar Infrastructure as Media of Resistance, or, Indigenous Solarities against Settler Colonialism." *South Atlantic Quarterly* 120, no. 1 (2021): 63–76.

King, Dean. *Skeletons on the Zahara: A Spectacular Odyssey, an Unforgettable Tale of Survival, Courage and Brotherhood*. London: Arrow Books, 2004.

Kirshner, Joshua, Vanesa Castán Broto, and Idalina Baptista. "Energy Landscapes in Mozambique: The Role of the Extractive Industries in a Post-conflict Environment." *Environment and Planning A: Economy and Space* 52, no. 1 (2020): 1051–71.

Klinger, Julie Michelle. *Rare Earth Frontiers*. Ithaca: Cornell University Press, 2017.

Knight, Judson. "Gil Eannes Passes the Point of No Return at Cape Bojador—and Inaugurates a New Era in Exploration." Encyclopedia.com. March 28, 2022. https://www.encyclopedia.com/science/encyclopedias-almanacs-transcripts-and-maps/gil-eannes-passes-point-no-return-cape-bojador-and-inaugurates-new-era-exploration.

Kohn, Margaret, and Kavita Reddy. "Colonialism." In *Stanford Encyclopedia of Philosophy*, edited by Edward N. Zalta. Stanford: Stanford University, 2017. https://plato.stanford.edu/entries/colonialism/.

Kooy, Michelle, and Karen Bakker. "Technologies of Government: Constituting Subjectivities, Spaces, and Infrastructures in Colonial and Contemporary Jakarta." *International Journal of Urban and Regional Research* 32, no. 2 (2008): 375–91.

Kuo, Sai-Hua. "From Solidarity to Antagonism: The Uses of the Second-Person Singular Pronoun in Chinese Political Discourse." *Talk and Text* 22, no. 1 (2002): 29–55.

Labban, Mazen. "Preempting Possibility: Critical Assessment of the IEA's World Energy Outlook 2010." *Development and Change* 43, no. 1 (2012): 375–93.

Laclau, Ernesto, and Chantal Mouffe. *Hegemony and Socialist Strategy: Towards a Radical Democratic Politics*. London: Verso, 1985.

Lakhal, Hamza. *A Destiny in Windblown Poetry*. Paris: L'Harmattan RASD, 2013.

———. "Fighting for a University: Interview with Saharawi Poet Hamza Lakhal." Interview by Joanna Allan. *Tadween Publishing*. December 16, 2015. https://tadweenpublishing.com/blogs/news/80473217-fighting-for-a-university-interview-with-saharawi-poet-hamza-lakhal.

Lakhal, Malainin. "Bedouin Obsession." *Exiled Ink!* 10, no. Autumn–Winter (2008): 25. http://www.exiledwriters.co.uk/magazines/em10.pdf.

Lalutte, Pauline. "Sahara: Toward an Analysis." *Middle East Research and Information Project Inc. (MERIP)* 45 (1976): 7–12.

Lawhon, Mary, David Nilsson, Jonathan Silver, Henrik Ernstson, and Shuaib Lwasa. "Thinking Through Heterogeneous Infrastructure Configurations." *Urban Studies* 55, no. 4 (2018): 720–32.

Lemanski, Charlotte. "Infrastructural Citizenship: The Everyday Citizenships of Adapting and/or Destroying Public Infrastructure in Cape Town, South Africa." *Transactions of the British Institute of Geographers* 45, no. 3 (2020): 589–605.

Lennon, Myles. "Energy Transitions in a Time of Intersecting Precarities: From Reductive Environmentalism to Antiracist Praxis." *Energy Research & Social Science* 73 (2021): 1–10.

Leyden, John, and Hugh Murray. *Historical Account of Discoveries and Travels in Africa, Volume I*. Edinburgh: Archibald Constable and Company and Longman, Hurst, Rees, Orme and Brown, 1818.

Livingstone, David N. "Darwinian Hippocratics, Eugenic Enticements, and the Biometeorological Body." In *Weather, Climate, and the Geographical Imagination*, edited by Martin Mahony and Samuel Randalls, 191–214. Pittsburgh: University of Pittsburgh Press, 2020.

Lloyd, David, and Patrick Wolfe. "Settler Colonial Logics and the Neoliberal Regime." *Settler Colonial Studies* 6, no. 2 (2016): 109–18.

Lohmann, Larry. "El cuestionamiento de la transición energética." *Boletín ECOS* 33 (November 2015–February 2016): 1–8.

———. "White Climate, White Energy." 2020. Draft for a forthcoming forum in Social Anthropology.

López, Elsa. "El viento como metáfora de la locura." Canarias Ahora. Accessed June 22, 2022. https://www.eldiario.es/canariasahora/lapalmaahora/sociedad/el-viento-como-metafora-de-la-locura-elsa-lopez_1_4488571.html.

Lucas, Joana. "Orientalism and Imperialism in French West Africa, Considerations on Travel Literature, Colonial Tourism, and the Desert as 'Commodity' in Mauritania." In *Tourism in the Global South: Heritages, Identities and Development*, edited by João Sarmento and Eduardo Brito-Henriques, 25–43. Lisbon: Centre for Geographical Studies, University of Lisbon, 2013.

Lucente, Adam. "Emirati Firm to Build Solar Power Plants in Morocco." Al Monitor. April 20, 2022). https://www.al-monitor.com/originals/2022/04/emirati-firm-build-solar-power-plants-morocco.

Luque-Ayala Andrés, and Jonathan Silver, ed. *Energy, Power, and Protest on the Urban Grid: Geographies of the Electric City*. London: Routledge, 2016.

Lydon, Ghislaine. "Writing Trans-Saharan History: Methods, Sources and Interpretations across the African Divide." In *The Sahara: Past, Present and Future*, edited by Jeremy Keenan. London: Routledge, 2007.

Macdonald, Graeme. "Containing Oil: The Pipeline in Petroculture." In *Petrocultures: Oil, Politics, Culture*, edited by Sheena Wilson, Adam Carlson, and Imre Szeman, 36–77. Montreal: McGill-Queen's University Press, 2017.

———. "Oil and World Literature." *American Book Review* 33, no. 3 (2012): 7–31.

Mahony, Martin, and Georgina Endfield. "Climate and Colonialism." *WIREs Climate Change* 9, no. 510 (2018): 1–16.

Márquez, Xavier. *Non-democratic Politics: Authoritarianism, Dictatorship, and Democratisation*. London: Red Globe Press, 2016.

Martín Beristain, Carlos, and Eloísa González Hidalgo. *El oasis de la memoria: Memoria histórica y violaciones de derechos en el Sáhara occidental: Tomo I, II*. Biscay: Institute of International Development and Cooperation, University of the Basque Country, 2012.

Martínez Milán, Jose María. "Empresa pública y minería en el Sahara occidental: Fosfatos de Bu Craa, S. A., 1969–1983." *Boletín geológico y minero* 128, no. 4 (2017): 913–29.

Mastini, Riccardo. "Degrowth: The Case for a New Economic Paradigm." Open Democracy. June 8, 2017. https://www.opendemocracy.net/en/degrowth-case-for-constructing-new-economic-paradigm/.

Mayrata, Ramón. *El imperio desierto*. Madrid: Calamar Ediciones, 2008.

Mbembe, Achille. "Necropolitics." *Public Culture* 15, no. 1 (2003): 11–40.

McClintock, Anne. *Imperial Leather: Race, Gender and Sexuality in the Colonial Contest*. New York: Routledge, 1995.

Mcfarlane, Colin, and John Rutherford. "Political Infrastructures: Governing and Experiencing the Fabric of the City." *International Journal of Urban and Regional Research* 32, no. 2 (2008): 363–74.

Merleau-Ponty, Maurice. *Phenomenology of Perception*. London: Routledge, 2012 (first published 1945).

Merriman, Hardy. "Theory and Dynamics of Nonviolent Action." In *Civilian Jihad: Nonviolent Struggle, Democratization, and Governance in the Middle East*, edited by Maria J. Stephan, 17–29. New York: Palgrave Macmillan, 2009.

Middle East Eye. "Morocco Bought Israel's Suicide Drones in $22m Arms Deal: Report." Accessed June 23, 2022. https://www.middleeasteye.net/news/israel-morocco-arms-deal-signed-suicide-drone.

Middle East Monitor. "Morocco, Israel to Build 2 Drone Factories." Accessed June 23, 2022. https://www.middleeastmonitor.com/20211217-morocco-israel-to-build-2-drone-factories/.

Mitchell, Timothy. *Carbon Democracy: Political Power in the Age of Oil*. New York: Verso, 2013.

Mohammed Salam, Fatma Galia. *Nada es eterno*. Bilboa: Lankopi, 2010.

Monod, Théodore. *Les Rosas de Sancta Marya de Gil Eanes (1434)*. Vol. 107. Coimbra: Centro de Estudos de Cartografia Antiga. Secçao de Coimbra, 1978.

Moore, Jason W. "Introduction: World-Ecological Imaginations." *Review: World Ecological Imaginations* 37, no. 3–4 (2014): 165–72.

Moustakbal, Jawal. *The Moroccan Energy Sector: A Permanent Dependence*. Amsterdam: Transnational Institute, 2021. https://longreads.tni.org/the-moroccan-energy-sector.

Moya, Conx. "La cuestión saharaui y su recepción en la literatura española." *Haz lo que debas* (blog). September 24, 2011. http://hazloquedebas.blogspot.com/2011/10/la-cuestion-saharaui-y-su-recepcion-en.html.

Mrázek, Rudolf. *Engineers of Happy Land: Technology and Nationalism in a Colony*. Princeton, NJ: Princeton University Press, 2002.

Mulero Clemente, Manuel. *Los territorios españoles del Sahara y sus grupos nómadas*. Las Palmas de Gran Canaria: El Siglo, 1945.

Narbón, Alfredo E. *Tierra seca*. Madrid: Editorial Bitacora, 1992.

Ngounou, Boris. "Morocco: Sharing Experience in Renewable Energy with Africa." Afrik 21. December 20, 2018. https://www.afrik21.africa/en/morocco-sharing-experience-in-renewable-energy-with-africa/.

Niblett, Michael. "Oil on Sugar: Commodity Frontiers and Peripheral Aesthetics." In *Global Ecologies and the Environmental Humanities: Postcolonial Approaches*, edited by Elizabeth DeLoughrey, Jill Didur, and Anthony Carrigan. Abingdon: Routledge, 2015.

Nixon, Rob. *Slow Violence and the Environmentalism of the Poor*. Cambridge, MA: Harvard University Press, 2011.

North Africa Post. "Morocco Reaps Diplomatic Gains of Soft Power in Africa." April 10, 2019. https://northafricapost.com/29771-morocco-reaps-diplomatic-gains-of-soft-power-in-africa.html.

Nye, David E. "Energy Narratives." *American Studies in Scandinavia* 25 (1993): 73–91.

Nye, Joseph S. *Hard, Soft and Smart Power: The Oxford Handbook of Modern Diplomacy*. Oxford: Oxford University Press, 2013.

Observatorio de derechos humanos y empresas en el Mediterráneo. *Los tentáculos de la ocupación: Informe sobre la explotación de los recursos pesqueros del Sáhara occidental en el marco de la ocupación del Estado Marroquí*. Barcelona: Observatorio de derechos humanos y empresas en el Mediterráneo, 2019. https://www.polisarioeuskadi.eus/wp-content/uploads/2019/05/TENTACULOS_OCUPACION_ESP_WEB1.pdf.

Oceanic Développement. *Evaluation ex-post du Protocole actuel d'Accord de partenariat dans le domaine de la Peche entre l'Union Europeenne et le Royaume de Maroc, etude d'impact d'un possible futur Protocole d'Accord*. Brussels: European Commission, 2010. https://www.fishelsewhere.eu/files/dated/2012-03-05/evaluation-app-maroc-2010.pdf.

OCP Group. *2019 Sustainability Report*. 2020. Accessed December 27, 2023. https://ocpsiteprodsa.blob.core.windows.net/media/2021-06/SUSTAINABILITY_report_2019.pdf.

Odartey-Wellington, Dorothy. "Walls, Border, and Fences in Hispano-Saharawi Creative Expression." *Research in African Literatures* 48, no. 3 (2017): 1–17.

Oldschool, Oliver. "Dangers of the West Coast of Africa: Shipwreck of the Sophia." *The Portfolio* 15 (1823): 1–21.

ONEE. *Onee au Maroc at en Afrique: Activité électricité*. Casablanca: ONEE, 2016. http://www.one.org.ma/FR/pdf/Brochure_ONEE_Africa_VF_COP22_V2.pdf.

Paddock, Judah. *A Narrative of the Shipwreck of the Oswego, on the Coast of South Barbary, and of the Sufferings of the Master and the Crew While in Bondage among the Arabs; Interdispersed with Numerous Remarks upon the Country and Its Inhabitants and the Peculiar Perils of That Coast*. London: Longman, Hurst, Rees, Orme, and Brown, 1818.

Palmer, Geraldine L., Jesica Siham Fernández, Gordon Lee, Hana Masud, Sonja Hilson, Catalina Tang, Dominique Thomas, Latriece Clark, Bianca Guzman, and Ireri Bernai. "Oppression and Power." In *Introduction to Community Psychology: Becoming an Agent of Change*, edited by Leonard A. Jason, Olya Glantsman, Jack F. O'Brien, and Kaitlyn N. Ramian. Montreal: Rebus Community, 2019.

Petrocultures Research Group. *After Oil*. Alberta: Petrocultures Research Group, 2016.

Ponzanesi, Sandra. *Paradoxes of Postcolonial Culture: Contemporary Women Writers of the Indian and Afro-Italian Diaspora*. New York: State University of New York Press, 2004.

Porges, Matthew. "Environmental Challenges and Local Strategies in Western Sahara." *Forced Migration Review: Climate Crisis and Local Communities* 64 (2020): 7–10.

Porges, Matthew, and Christian Leuprecht. "The Puzzle of Nonviolence in Western Sahara." *Democracy and Security* 12, no. 2 (2016): 65–84.

Power, Marcus, and Joshua Kirshner. "Powering the State: The Political Geographies of Electrification in Mozambique." *Environment and Planning C: Politics and Space* 37, no. 3 (2019): 498–518.

Power, Marcus, Peter Newell, Lucy Baker, Harriet Bulkeley, Joshua Kirshner, and Adrian Smith. "The Political Economy of Energy Transitions in Mozambique and South Africa: The Role of the Rising Powers." *Energy Research and Social Science* 17 (2016): 10–19.

Quijano, Aníbal. "Coloniality of Power, Eurocentrism, and Latin America." *Nepantla: Views from South* 1, no. 3 (2000): 533–80.

Quiroga, Francisco. "Expedición al Sáhara." *Revista de geografía comercial* 25–30 (1886): 1–110.

Rankin, William J. "Infrastructure and the International Governance of Economic Development, 1950–1965." In *Internationalization of Infrastructures*, edited by Jean-Francois Auger, Jan Jaap Bouma, and Rolf Künneke, 61–75. Delft: Delft University of Technology, 2009.

Reverte, Javier. *El médico de Ifni*. Barcelona: Debolsillo, 2007 (first published 2005).

Riley, James. *Sufferings in Africa, Captain Riley's Narrative: An Authentic Narrative of the Loss of the American Brig Commerce*. New York: Lyons Press, 2000.

———. *Sufferings in Africa: The Incredible True Story of a Shipwreck, Enslavement, and Survival on the Sahara*. London: John Murray, 1817.

Robbins, Archibald. *A Journal, Comprising an Account of the Loss of the Brig Commerce*. Hartford: S. Andrus, 1848.

Rodríguez, Monica. *Sand/Water*. Zaragoza: Bubisher, 2019.

Rodriguez Esteban, José Antonio. "El mapa del África Occidental Española de 1949 a escala 1:500.000: Orgullo militar, camelladas y juegos poéticos saharauis." *Cybergeo: European Journal of Geography* (2011): document 516. http://journals.openedition.org/cybergeo/23461.

Rogers, Douglas. "Energopolitical Russia: Corporation, State, and the Rise of Social and Cultural Projects." *Anthropological Quarterly* 87, no. 2 (2014): 431–51.

Romeo, Félix. *Discothèque*. Barcelona: Editorial Anagrama, 2001.

Ruano Posada, Violeta. "The Sahara Is Not for Sale: The Song of a Revolution." *Violeta Ruano Music* (blog). May 19, 2014. http://violetaruanomusic.blogspot.com/2014/05/the-sahara-is-not-for-sale-song-of.html.

Ruano Posada, Violeta, and Vivian Solana Moreno. "The Strategy of Style: Music, Struggle, and the Aesthetics of Sahrawi Nationalism in Exile." *Transmodernity: Journal of Peripheral Cultural Production of the Luso-Hispanic Word* 5, no. 3 (2015): 40–64.
Ruiz, Fico. "Emilio Bonelli, el creador del Sáhara español." *Aragonautas: Aragoneses olvidados. Náufragos de la historia* (blog). November 6, 2013. http://aragonautas.blogspot.com/2013/11/emilio-bonelli-el-creador-del-sahara.html.
Rupp, Stephanie. "Considering Energy: E = Mc2 = (Magic X Culture)2." In *Cultures of Energy: Power, Practices, Technologies*, edited by Sarah Strauss, Stephanie Rupp, and Thomas Love, 79–95. Walnut Creek, CA: Left Coast Press, 2013.
Saharawi Arab Democratic Republic. *First Indicative Nationally Determined Contribution*. Rabouni: Office of the Prime Minister, 2021. https://www.spsrasd.info/news/sites/default/files/documents/sadr_ndc_draft_final_09nov21.pdf.
Saharawi Natural Resource Watch. *Sahrawis: Poor People in a Rich Country*. Tindouf: Saharawi Natural Resource Watch, 2013. https://issuu.com/moulud/stacks/fc4a1bf5ef9841a397b223311bc6c7f5.
Saharawis. *Petition of Signatures Addressed to the General Governor of the Sahara Province*. El Aaiún, September 14, 1974.
Said, Edward. *Orientalism*. London: Penguin, 1995.
Saint-Exupéry, Antoine de. *Wind, Sand and Stars*. London: Penguin, 2000 (first published 1939).
Sampedro Vizcaya, Benita. "The Colonial Politics of Meteorology: The West African Expedition of the Urquiola Sisters." In *Unsettling Colonialism: Gender and Race in the Nineteenth-Century Global Hispanic World*, edited by N. Michelle Murray and Akiko Tsuchiya, 19–54. New York: SUNY Press, 2019.
San Martín, Pablo. *Western Sahara: The Refugee Nation*. Cardiff: University of Wales Press, 2010.
San Martín, Pablo, and Joanna Allan. *The Largest Prison in the World: Landmines, Walls, UXOs and the UN's Role in the Western Sahara*. Madrid: Spanish Strategic Studies Group (GEES), 2007. https://researchportal.northumbria.ac.uk/ws/portalfiles/portal/42975238/Landmines_and_UXOs_in_Western_Sahara_libre.pdf.
San Martín, Pablo, and Ben Bollig, eds. *Treinta y uno—Thirty One: A Bilingual Anthology of Saharawi Resistance Poetry in Spanish*. London: Ediciones Sombrerete, 2007.
Saugnier, F., and Pierre-Raymond Brisson. *Voyages to the Coast of Africa by Mess. Saugnier and Brisson, Containing an Account of Their Shipwreck on Board Different Vessels, and Subsequent Slavery, and Interesting Details of the Manners of the Arabs of the Desert*. London: G. G. J. and J. Robinson, 1792.
Schivelbusch, Wolfgang. *Disenchanted Night: The Industrialization of Light in the Nineteenth Century*. Berkeley: University of California Press, 1988.
Schmidt, Leigh Erik. *Hearing Things: Religion, Illusion, and the American Enlightenment*. Cambridge, MA: Harvard University Press, 2000.
Scholz, Sally J. *Political Solidarity*. University Park: Pennsylvania University Press, 2008.
Schwarz, Suzanne. "Introduction." In *Slave Captain: The Career of James Irving in the Liverpool Slave Trade*, edited by Suzanne Schwarz, 9–81. Wrexham, Clwyd: Bridge Books, 1995.
Scott, Alexander. "Article X: Account of Alexander Scott, Who Spent Some Years in the Interior of Africa." *Annals of Philosophy* 11 (1818): 122–24.
Sears, Christine E. *American Slaves and African Masters: Algiers and the Western Sahara, 1776–1820*. New York: Palgrave Macmillan, 2012.
Seth, Susan. "Putting Knowledge in Its Place: Science, Colonialism, and the Postcolonial." *Postcolonial Studies* 12, no. 4 (2009): 373–88.
Shaheed, Farida. *Report of the Independent Expert in the Field of Cultural Rights*. United Nations: United Nations Human Rights Council, 2012. https://digitallibrary.un.org/record/729775?ln=en.

Shalhoub-Kevorkian, Nadera. "The Occupation of the Senses: The Prosthetic and Aesthetic of State Terror." *British Journal of Criminology* 57, no. 6 (2017): 1279–300.

Sharp, Gene. *How Nonviolent Struggle Works*. East Boston: Albert Einstein Institute, 2013.

Short, Damien. *Redefining Genocide: Settler Colonialism, Social Death and Ecocide*. London: Zed Books, 2016.

Siemens. "About Us." Accessed May 19, 2019. https://www.siemensgamesa.com/en-int/about-us.

———. *Pictures of the Future: The Magazine for Research and Innovation*. Munich: Siemens, 2010.

———. "Supporting Morocco's Sustainable Development for a Brighter Future for the Entire Country." April 18, 2019. http://new.siemens.com/ma/en/company/topic-areas/ingenuity-for-life.html.

———. "Turning the Wind's Energy into a Source of Economic Development: That's Ingenuity for Life." March 20, 2019. https://www.siemens.com/ma/en/home/company/topic-areas/ingenuity-for-life/tarfaya-wind-farm.html.

Silver, Jonathan. "Incremental Infrastructures: Material Improvisation and Social Collaboration across Post-Colonial Accra." *Urban Geography* 35, no. 6 (2014): 788–804.

Snader, Joe. "The Oriental Captivity Narrative and Early English Fiction." *Eighteenth Century Fiction* 9, no. 3 (1997): 267–98.

Sorenson, Janet. "Logging-In: The Ship's Log as Medium." Arcade. 2015. Accessed December 27, 2023. https://arcade.stanford.edu/content/logging-ships-log-medium-0.

Sørfonn Moe, Tone. *Observer Report: The 2017 Trial against Political Prisoners from Western Sahara*. September 21, 2017. https://papers.ssrn.com/sol3/papers.cfm?abstract_id=3050803.

Sörlin, Sverker, and Melissa Lane, "Historicizing Climate Change—Engaging New Approaches to Climate and History." *Climatic Change* 151 (2018): 1–13.

Spice, Anne. "Fighting Invasive Infrastructures: Indigenous Relations against Pipelines." *Environment and Society* 9 (2018): 40–56.

Steinman, Erich W. "Decolonization Not Inclusion: Indigenous Resistance to American Settler Colonialism." *Sociology of Race and Ethnicity* 2, no. 2 (2016): 219–36.

Stephan, Maria J., and Jacob Mundy. "A Battlefield Transformed: From Guerrilla Resistance to Mass Non-violent Struggle in Western Sahara." *Journal of Military and Strategic Studies* 8, no. 3 (2006): 1–32.

Swanson, Ana, and Christopher Buckley. "Chinese Solar Companies Reportedly Linked to Forced Labour Allegations in Xinjiang." *New York Times*, January 8, 2021. https://www.business-humanrights.org/en/latest-news/china-solar-companies-linked-to-forced-labour-allegations/.

Szeman, Imre. "Conjectures on World Energy Literature: Or, What Is Petroculture?" *Journal of Postcolonial Writing* 53, no. 3 (2017): 278.

Szeman, Imre and Darin Barney. "From Solar to Solarity." *South Atlantic Quarterly* 120, no. 1 (2021): 1–11.

Szeman, Imre, Ruth Beer, Warren Cariou, Mark Simpson, Sheena Wilson, Darren Fleet, Jordan Kinder, and Michael Minor. *On the Energy Humanities: Contributions from the Humanities, Social Sciences, and Arts to Understanding Energy Transition and Energy Impasse*. Alberta: University of Alberta, 2016.

Szeman, Imre, and Dominic Boyer. "Introduction." In *Energy Humanities: An Anthology*, edited by Imre Szeman and Dominic Boyer. Baltimore: John Hopkins University Press, 2017.

Taine-Cheikh, Catherine. "Following the Tracks of the Moors, between Sand and Dust." *Techniques & Culture: Revue semestrielle d'anthropologie des techniques* 61 (2013): 188–209.

Taine-Cheikh, Catherine. "Ḥassāniyya Arabic in Contact with Berber: The Case of Quadriliteral Verbs." In *Arabic in Contact: Studies in Arabic Linguistics*, edited by Stefano Manfredi and Mauro Tosco, 135–60. Amsterdam: John Benjamins Publishing Company, 2018.

Tavakoli, Judit. "Cultural Entrepreneurship of Sahrawi Refugees." *African Identities* 18, no. 3 (2020): 279–94.

Technical director of Díaz de Villegas power station. "Solicitud viviendas del I.N.V." January 15, 1966.

Thompson, Carl. "Introduction." In *Shipwreck in Art and Literature: Images and Interpretations from Antiquity to the Present Day*, edited by Carl Thompson. New York: Routledge, 2013.

Together to Remove the Wall. "Impacts of the Wall." Accessed December 27, 2023. http://removethewall.org/the-wall/impacts-of-the-wall/.

Torbado, Jesús. *El imperio de arena*. Barcelona: Plaza & Janés Editores, 1998.

Tuan, Yi-Fu. "Space and Place: Humanistic Perspective." In *Philosophy in Geography (Theory and Decision Library)*, edited by Gale Stephen and Gunnar Olsson. Dordrecht: Springer, 1979.

Tuck, Eve, and K. Wayne Yang. "Decolonization Is Not a Metaphor." *Decolonization: Indigeneity, Education and Society* 1, no. 1 (2012): 1–40.

United Nations General Assembly. *Resolution 35/19: The Question of Western Sahara*. November 11, 1980. http://www.securitycouncilreport.org/atf/cf/%7B65BFCF9B-6D27-4E9C-8CD3-CF6E4FF96FF9%7D/a_res_35_19.pdf.

van Gelder, Geert Jan. "The Abstracted Self in Arabic Poetry." *Journal of Arabic Literature* 14 (1983): 22–30.

van Son, Paul. "Desertec Industries—Enabling Desertec in Eumena." Siemens Sustainability Conference, Istanbul. June 1, 2010.

Vázquez-Figueroa, Alberto. *Arena y viento*. Barcelona: Delbolsillo, 2005 (first published 1961).

Veale, Lucy, and Georgina Endfield. "The Helm Wind of Cross Fell." *Weather* 69, no. 1 (2014): 3–7.

Veracini, Lorenzo. "Introducing: Settler Colonial Studies." *Settler Colonial Studies* 1, no. 1 (2011): 1–12.

———. "Settler Colonialism and Decolonisation." *Borderlands* 6, no. 2 (2007): 1–11.

———. *Settler Colonialism: A Theoretical Overview*. Basingstoke, UK: Palgrave Macmillan, 2010.

Verdeil, Eric. "Beirut, Metropolis of Darkness: The Politics of Urban Electricity Grids." In *Energy, Power, and Protest on the Urban Grid: Geographies of the Electric City*, edited by Andrés Luque-Ayala and Jonathan Silver, 155–75. London: Routledge, 2016.

Verweijen, Judith, and Alexander Dunlap. "The Evolving Techniques of Social Engineering, Land Control and Managing Protest against Extractivism: Introducing Political (Re)actions 'from Above.'" *Political Geography* 83 (2021): 1–9.

Viswanathan, Leela. "Indigenous Engagement with Renewable Energy Projects." Indigenous Climate Hub. June 16, 2021. https://indigenousclimatehub.ca/2021/06/indigenous-engagement-with-renewable-energy-projects/.

Volpato, Gabriele, and Patricia Howard. "The Material and Cultural Recovery of Camels and Camel Husbandry among Sahrawi Refugees of Western Sahara." *Pastoralism* 4, no. 7 (2014): 1–23.

Volpato, Gabriele, and Rajindra Kumar Puri. "Dormancy and Revitalization: The Fate of Ethnobotanical Knowledge of Camel Forage among Sahrawi Nomads and Refugees of Western Sahara." *Ethnobotany Research and Applications* 12 (2014): 183–210.

Watson, Lyall. *Heaven's Breath: A Natural History of the Wind*. London: Hodder and Stoughton, 1984.

Weexsteeny, Raoul. "Fighters in the Desert." *MERIP Reports* 45 (1976): 3–5.

Wenzel, Jennifer. *The Disposition of Nature: Environmental Crisis and World Literature*. New York: Fordham University Press, 2019.

Western Sahara Resource Watch. "Court Annuls EU Deals in Occupied Western Sahara." September 29, 2021. https://wsrw.org/en/news/court-orders-halt-to-eu-deals-in-occupied-ws.

———. *Dirty Green March: Morocco's Controversial Renewable Energy Projects in Occupied Western Sahara*. Brussels: Western Sahara Resource Watch and Emmaus Stockholm, 2013. https://wsrw.org/files/dated/2013-08-27/dirty_green_march_eng.pdf.

———. *Fuelling the Occupation: The Swedish Transport of Oil to Occupied Western Sahara*. Stockholm: Emmaus Stockholm and Western Sahara Resource Watch, 2014. https://wsrw.org/files/dated/2014-06-20/fuelling_the_occupation_2014_web.pdf.

———. *Greenwashing Occupation: How Morocco's Renewable Energy Projects in Occupied Western Sahara Prolong the Conflict over the Last Colony in Africa*. Brussels: Western Sahara Resource Watch, 2021. https://wsrw.org/en/news/report-morocco-uses-green-energy-to-embellish-its-occupation.

———. "Hunger Striking against EU Fisheries in Western Sahara." June 8, 2011. https://wsrw.org/en/archive/1953.

———. *Label and Liability*. Stockholm: Western Sahara Resource Watch and Emmaus Stockholm, 2012. https://wsrw.org/files/dated/2012-06-17/wsrw_labelliability_2012.pdf.

———. "Morocco Continues to Discard by-Catches in Occupied Western Sahara." 2015. Accessed December 27, 2023. https://www.fishelsewhere.eu/a140x1484.

———. "Saharawi Family Home Lost to Moroccan Power Line." October 22, 2014. https://www.wsrw.org/a106x2998.

———. "Saharawis: Check Out Your Fish Here." November 19, 2013. https://wsrw.org/en/archive/2712.

———. "Self-Immolation by Moroccan Sea Captain in Dakhla." April 11, 2017. https://wsrw.org/en/archive/3802.

———. "Siemens with Press Release on Project in Occupied Western Sahara." January 28, 2013. https://wsrw.org/en/archive/2494.

———. *Totally Wrong: Total Sa in Occupied Western Sahara*. Brussels: Western Sahara Resource Watch, 2013. https://emmausstockholm.se/wp-content/uploads/2017/11/totally_wrong.pdf.

———. "Trawler in Occupied Western Sahara Involved in Massive Discards." May 7, 2016. https://wsrw.org/en/archive/3484.

———. "UK Takes Lead in Gas Exports to Occupied Western Sahara." June 2, 2021. https://wsrw.org/en/news/uk-takes-lead-in-gas-exports-to-occupied-western-sahara.

White, Tom. "Climate, Power, and Possible Futures, from the Banks of the Humber Estuary." *Open Library of Humanities* 5, no. 1 (2019): 1–35.

Williams, Rhys. "'This Shining Confluence of Magic and Technology': Solarpunk, Energy Imaginaries, and the Infrastructures of Solarity." *Open Library of Humanities* 5, no. 1 (2019): 1–35.

Wilson, Alice. *Sovereignty in Exile: A Saharan Liberation Movement Governs*. Philadelphia: University of Pennsylvania Press, 2016.

Wilson, Sheena, Imre Szeman, and Adam Carlson. "On Petrocultures: Or, Why We Need to Understand Oil to Understand Everything Else." In *Petrocultures: Oil, Politics, Culture*, edited by Sheena Wilson, Imre Szeman, and Adam Carlson, 3–20. Montreal: McGill-Queen's University Press, 2017.

Winther, Tanja. *The Impact of Electricity: Development, Desires, and Dilemmas*. Oxford: Berghahn, 2008.

Winther, Tanja, and Harold Wilhite. "Tentacles of Modernity: Why Electricity Needs Anthropology." *Journal of Cultural Anthropology* 30, no. 4 (2015): 569–77.

Wolfe, Patrick. *Settler Colonialism and the Transformation of Anthropology: The Politics and Poetics of an Ethnographic Event*. London: Bloomsbury Publishing, 1999.

Yang, Yu-You, Huai-Na Wu, Shui-Long Shen, Suksun Horpibulsuk, Ye-Shuang Xu, and Qing-Hong Zhou. "Environmental Impacts Caused by Phosphate Mining and Ecological Restoration: A Case History in Kunming, China." *Hazards* 74 (2014): 755–70.

Yu, Hongbin, Mian Chin, Tianle Yuan, Huisheng Bian, Lorraine A. Remer, Joseph M. Prospero, Ali Omar, et al. "The Fertilizing Role of African Dust in the Amazon Rainforest: A First Multiyear Assessment Based on Calipso Lidar Observations." *Geophysical Research Letters* 42 (2015): 1984–91.

Yuval-Davis, Nira. *Gender and Nation*. London: Sage Publications, 1997.

Zafirovski, Milan. *The Protestant Ethic and the Spirit of Authoritarianism: Puritanism, Democracy, and Society*. New York: Springer, 2007.

Zaman, Saifuz Md. "The Edification of Sir Walter Scott's Saladin in the Talisman." *Studies in Literature and Language* 1, no. 8 (2010): 39–46.

Zanon, Bruno. "Infrastructure Network Development, Re-territorialization Processes and Multilevel Territorial Governance: A Case Study in Northern Italy." *Planning Practice and Research* 26, no. 3 (2011): 325–47.

Zunes, Stephen, and Jacob Mundy. *Western Sahara: War, Nationalism, and Conflict Irresolution*. New York: Syracuse University Press, 2010.

Index